Spatial Statistical Methods for Geography

Sara Miller McCune founded SAGE Publishing in 1965 to support the dissemination of usable knowledge and educate a global community. SAGE publishes more than 1000 journals and over 800 new books each year, spanning a wide range of subject areas. Our growing selection of library products includes archives, data, case studies and video. SAGE remains majority owned by our founder and after her lifetime will become owned by a charitable trust that secures the company's continued independence.

Los Angeles | London | New Delhi | Singapore | Washington DC | Melbourne

Spatial Statistical Methods for Geography

Peter A. Rogerson

Los Angeles | London | New Delhi
Singapore | Washington DC | Melbourne

Los Angeles | London | New Delhi
Singapore | Washington DC | Melbourne

SAGE Publications Ltd
1 Oliver's Yard
55 City Road
London EC1Y 1SP

SAGE Publications Inc.
2455 Teller Road
Thousand Oaks, California 91320

SAGE Publications India Pvt Ltd
B 1/I 1 Mohan Cooperative Industrial Area
Mathura Road
New Delhi 110 044

SAGE Publications Asia-Pacific Pte Ltd
3 Church Street
#10-04 Samsung Hub
Singapore 049483

Editor: Jai Seaman
Senior assistant editor: Charlotte Bush
Assistant editor, digital: Sunita Patel
Production editor: Katherine Haw
Copyeditor: Catja Pafort
Proofreader: Neville Hankins
Indexer: Martin Hargreaves
Marketing manager: Ben Sherwood
Cover design: Shaun Mercier
Typeset by: C&M Digitals (P) Ltd, Chennai, India

Library of Congress Control Number: 2020944349

British Library Cataloguing in Publication data

A catalogue record for this book is available from the British Library

ISBN 978-1-5297-0745-8
ISBN 978-1-5297-0744-1 (pbk)

Contents

List of Figures

List of Tables

About the Author

Peter A. Rogerson is SUNY (State University of New York) Distinguished Professor in the Department of Geography at the University at Buffalo, Buffalo, New York, USA. He also holds an adjunct appointment in the Department of Biostatistics.

Acknowledgements

I would like to acknowledge the helpful comments of Peter Kedron, Xiang Ye, and the students of my graduate class in Spatial Statistics. I would also like to acknowledge the support and very helpful comments from Charlotte Bush, Katherine Haw, Catja Pafort, Robert Rojek, and Jai Seaman at SAGE Publishing.

Preface

Statistical methods for spatial data have advanced significantly during the last few decades. There is a wealth of research, writing, and information on spatial statistics, spatial modeling, and the estimation of the parameters of spatial models. As the fields of spatial analysis and spatial econometrics have matured, many highly sophisticated explications of spatial processes and methods have emerged. This is mostly good news, because it reflects the tremendous progress that has been made. But one might argue that the steps forward have been so large and so quick that they have not been communicated effectively to audiences at all levels. In particular, there exists a fairly significant gap between treatments of the subject matter and the ability of, say, (even advanced) graduate students to understand them. A related side effect of the rapid growth is that there is also a sizable gap between the increasing sophistication of software packages that implement methods and the ability of many to understand their details. Thus to a large extent, these packages are black boxes.

A major objective here is to bridge these perceived gaps. I have attempted to do this by writing a book at a level that many students will find themselves after one or two courses in statistics. What is needed is a text, aimed primarily at students, that has the objective of introducing spatial statistical methods in a gentle manner. That is the chief aim of this book; it has been written from the viewpoint of building upon an introductory course in statistics, and it contains expository examples and elementary exercises. It should therefore provide a solid footing for the interested reader to then both apply and interpret the methods, and read and understand the more advanced texts.

The more general perspective is that there is now a two-volume set, with *Statistical Methods for Geography* (SMfG; currently in its 5th edition) serving as the first volume. The importance of spatial methods is indicated there, but the details and primary material are in this current book, *Spatial Statistical Methods for Geography* (SSMfG). The primary focus here is on (a) methods of geographical cluster detection, and (b) the extension of traditional statistical methods for use with spatial data. With regard to the former, many of the topics in *Statistical Detection and Surveillance of Geographic Clusters* (which I coauthored with Ikuho Yamada) are also covered here, but the exposition here is simpler, with elementary examples and exercises, and it is aimed at students rather than researchers.

The coverage of material and expositional style follow, for the most part, that which might be expected for a textbook. As already indicated, the overriding aim is to bridge the gap between treatments of elementary statistics and books that address advanced spatial statistics. There are a number of good books on applied spatial statistics that have recently become available, but the intent here is to solidify conceptual foundations. In most places within the text, this means that basic concepts are introduced, explained, and illustrated.

This book is written from the viewpoint that it is helpful to learn some of the concepts and methods of spatial statistics through seeing and doing examples that are carried out primarily by hand. Certainly teaching this material requires both mastery and understanding of each step and each idea. In carrying out examples by hand where possible, the student is led to think more carefully about each of the steps, and this in turn provides additional exposure for the student to the underlying concepts. Although this approach carries with it the risk that one might simply learn "what goes where" in equations without seeing the broader ideas the first time through, it is often the case that repetition will increase the likelihood that ideas are captured along with the methodological details.

This is a view that is not shared by all – indeed, it is perhaps even a minority view. The availability of software has facilitated applications. This of course is an extremely welcome advance (in the early days of GIS, a real limitation was the lack of such software). Unless programming itself is involved in the learning process, however, it is relatively easy and common for software applications to remain black boxes. There is input, and there is output, but to understand the output thoroughly, one way or the other a strong grasp on foundational ideas is important.

Regarding scope, this book is best viewed as one that might accompany a second course in descriptive and inferential statistics, where there is a desire to utilize and advance those methods for problems that use spatial data. There are many topics in spatial statistics that are either not examined here, or are not covered in an exhaustive manner. Geostatistical methods, modeling of point processes, and more detailed coverage of spatial regression are but three such examples. Additional resources pertaining to these and other topics are provided in the epilogue.

A primary purpose of this text is to provide a bridge to other treatments of topics in spatial statistics by (a) covering additional mathematical tools, and (b) demonstrating how descriptive and inferential statistics may be used in a spatial setting.

The prerequisites for the material covered here consist primarily of material from a first-semester course in statistics. If you are familiar with z-tests, t-tests, hypothesis testing involving means and proportions, confidence intervals, and the normal, Poisson, and binomial distributions, then you are likely in a good position to learn the material in this book. This short list is meant to give the reader a rough idea of the assumed level of background – it is not necessarily comprehensive. Nor is it assumed that the level of knowledge that the reader has is necessarily deep – knowledge of the key background material here can easily be refreshed.

We will also introduce other tools that serve as prerequisites for the substantive material to follow. This serves the additional aim of extending preparation for reading more advanced papers and texts. You may or may not have been exposed to some of these tools (such as matrix algebra, maximum likelihood estimation, and simulation) but hopefully they are introduced here in a relatively gentle manner. These tools are discussed in detail in Appendix A. A very reasonable strategy for tackling the material in this book would be to cover the entirety of that appendix prior to covering the material in Chapter 2 and beyond.

Online Resources

Spatial Statistical Methods for Geography is accompanied by a range of online resources to support learning and teaching. Find them at: https://study.sagepub.com/rogerson spatialstatistics

- **Video introductions** give you an overview of the key topics covered in each chapter, and **video demonstrations** offer a step-by-step run through of techniques and software.
- **Exercises with solutions** enable you to test your understanding of key concepts and practice the application of statistical methods.
- **.html pages and R Notebooks** for select chapters demonstrate how to implement key techniques.
- Three sample **datasets** used in the text are ready to download in numerous formats, allowing you to develop your statistical skills at your own pace.
- **Roadmaps** through the book give examples of how you can combine the methods and materials in the book and apply them to different topics.
- A **topic-software chart** provides a list of software you can use to implement techniques discussed in each chapter.

Lecturers can login to access:

Select **additional exercises** and solutions, with accompanying .html files and R Notebooks, which can be used to check your students' understanding.

1

Introduction to Spatial Statistical Methods for Geography

Spatial dependence is a defining feature of geographic data. Observations that are close in space are likely to be correlated. If for example there are two respondents to a survey from the same household, their responses are likely to be similar. Two soil samples from nearby locations are likely to convey similar information. Effectively then, we shouldn't count these as two observations – they represent the equivalent of one, or perhaps slightly more than one, observation.

Therefore, in carrying out studies with geographic data, a fundamental assumption of statistical analysis – that of independent observations – is often violated. This of course has consequences for the interpretation of analyses. With this book, we pursue two main goals – to be able to detect statistically the existence of spatial patterns, and to be able to adjust either the methods of statistical analysis or our interpretation of standard statistical analyses when observations are not independent.

Both goals fall within the field of spatial statistical analysis. As we pursue these goals, we will build a basic foundation in some of the tools, methods, and concepts that will aid in making progress toward these goals. We will also touch upon other facets of spatial statistical analysis along the way, and will illustrate them with applications and exercises. For example, gaining knowledge of descriptive spatial statistics opens the door for some interesting questions and applications, and we will explore some of these. In some cases, these explorations will prove to be interesting from a substantive point of view, and they may (or may not!) be worth following up in more depth. In other cases, these explorations may prove useful for illustration, and, more importantly, for enhancing our abilities to work with valuable tools and methods.

There is also a third, and important, goal here. In addition to the substantive interests in correcting statistical analyses for spatially dependent observations and in detecting spatial patterns in data, the reader will hopefully extend their knowledge of probabilistic and statistical concepts to the point that thinking about questions in a statistical manner becomes quite natural. Through applications and extensions, and the introduction of material that may at times seem somewhat peripheral, the goal is to encourage the reader to first understand, and then begin to further examine, ways in which statistical thinking could be brought to bear on geographical questions.

1.1 Spatial Dependence

Traditional statistical methods such as difference-of-means tests and regression require different approaches when used with geographic data – this is because dependence among observations that are close in space reduces the effective size of the sample. Thus when you carry out statistical analyses on geographic data, the observations are not independent. This will affect the conclusions of our statistical analyses – we effectively have fewer observations than we think we do. Therefore we should not be as confident of our estimates, and we should not be so quick to reject null hypotheses. If such dependence is ignored, we will construct confidence intervals that are too narrow, and we will reject too many true null hypotheses. There is widespread recognition of these issues. It is also clear that an increasing proportion of journal articles now convey an awareness of the effects of spatial dependence. Many use spatial methods to address the presence of spatial dependence, but this is not universal. It is important to learn what the effects are, and what to do about them.

1.2 Spatial Patterns

A second objective is to recognize patterns in geographic data. We need to be able to look at maps and quantify deviations from spatial randomness – in part to be able to distinguish between what is random and what is not. When we look at a map of disease, how do we know if there is a raised risk somewhere? There may appear to be clusters of a few cases, but could they have arisen by chance alone?

The maps shown in Figure 1.1 depict kidney cancer mortality rates for US counties (Gelman and Nolan, 2017). One map shows the counties that are in the highest decile – that is, they are in the top 10% with respect to kidney cancer mortality rates.

The other map shows those counties in the lowest decile – these are the counties that have the country's lowest kidney cancer mortality rates. The maps have some similar patterning – in both maps, the Midwest and the Great Plains stand out. Some counties in these regions have the country's highest rates; others have the country's lowest rates.

Highest kidney cancer death rates

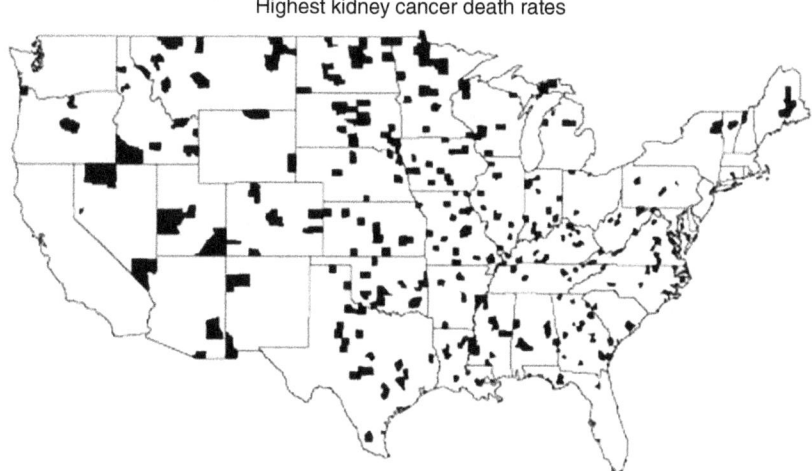

Figure 1.1a   Counties in the highest decile for kidney cancer mortality (1980–5)

Source: Gelman and Nolan (2017:14). Reproduced with permission of Oxford University Press.

Lowest kidney cancer death rates

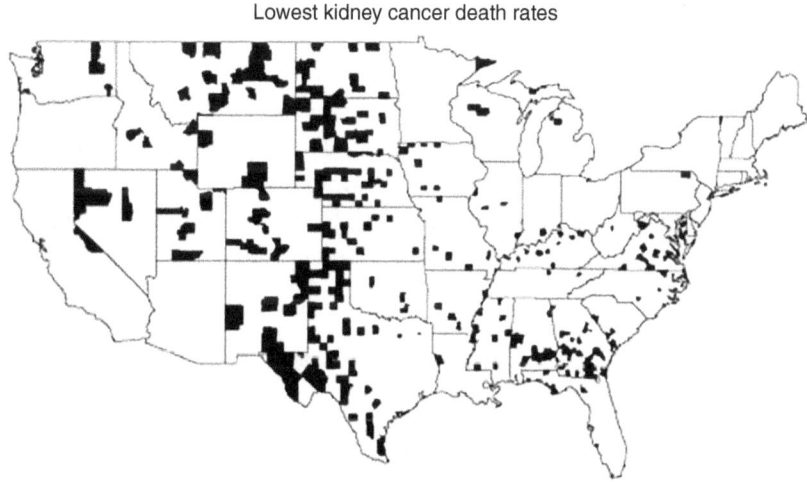

Figure 1.1b   Counties in the lowest decile for kidney cancer mortality (1980–9)

Source: Gelman and Nolan (2017:15). Reproduced with permission of Oxford University Press.

Thus the region with the lowest rates is the same as the region with the highest – both lowest and highest rates occur in the Midwest and the Great Plains.

Why might this be? The reason has to do with the low denominators used to compute the rates in this part of the country. A mortality rate is based upon the number of deaths, divided by the size of the population at risk of dying. In the Midwest and the Great Plains, population sizes (and hence the denominators of the rates) are relatively low. Just one case of kidney cancer could cause the rate to be high if the population was low enough. Similarly, if the population was very low, it would not be surprising to get an observed rate of zero. In low population counties then, rates are more variable.

A similar effect is observed when tossing coins. If you toss a coin ten times, rates at which the coin turns up heads are more variable than when you toss the coin 1,000 times. It would not be too surprising to get, say, anywhere from three to seven heads if you toss it ten times (i.e., rates of heads between 0.3 and 0.7 would not be surprising). But you certainly wouldn't expect this much variability if you were to toss the coin 1,000 times (the rate at which the coin came up heads, in repeated experiments, would be across a much narrower range (centered on 0.5) than the range from 0.3 to 0.7).

Simply stated, small samples display more variability than large samples. And we see that this has consequences for map reading. We need to be careful when we interpret maps of rates, because the places with the highest (and lowest) rates may have those rates not because their inherent risk is high (or low), but simply because they have small denominators.

1.3 Some Motivating Problems

1.3.1 Perceptions of Randomness – Visual Assessment of Maps

Part A

Figure 1.2 is a map of the locations of 20 cases of disease and the locations of 50 healthy controls. Suppose that you are a health analyst for a local agency and you are asked to find any region or regions that have more cases than would be expected, relative to the distribution of controls. Take the map (or a copy of it!) in Figure 1.2 and circle these regions.

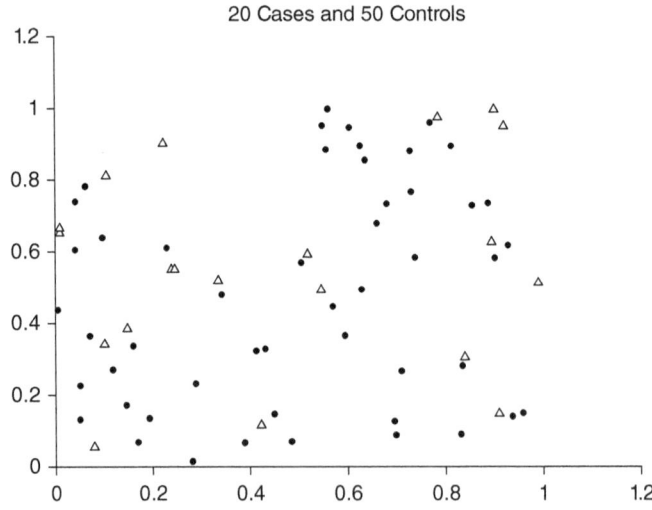

Figure 1.2   20 cases and 50 controls

Part B

The map in Figure 1.3 contains the locations of 50 healthy individuals drawn at random from a hypothetical population. This hypothetical population is randomly distributed throughout the region. Suppose that there are 20 individuals in the population who have a particular disease. Furthermore, we will assume that the disease risk is spatially uniform – there are no locations on the map that have a raised or suppressed risk of disease.

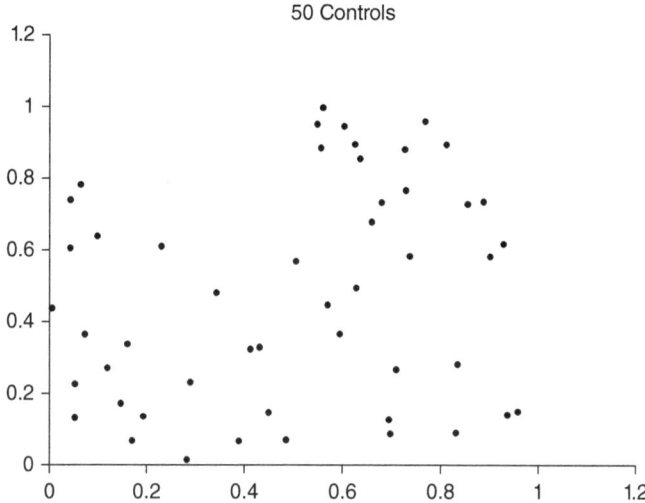

Figure 1.3   50 controls

Your task in this exercise is to place 20 points on the map, each representing the location of an individual with the disease. As you do so, keep in mind that the spatial distribution of the 20 points should not be influenced by the locations of the 50 healthy individuals (since the population is distributed randomly, and disease risk is spatially random). Thus the geographic distribution of cases should not be clustered or dispersed, relative to the distribution of healthy controls.

Discussion

The map in Part A was generated by choosing both the case and control locations at random. There is in fact no clustering of cases relative to controls, and thus nothing should have been circled. Of course a few cases may be near to one another just by chance. The human eye is good at organizing spatial information and "seeing" what it thinks are significant geographic clusters of points; the eye is not as good at distinguishing whether points that are close together could have arisen by chance alone.

For the exercise in Part B, look at the nearest point for each of the 20 cases on the map. In some instances the nearest point will be a case and in other instances it will be a control. Do this for each case, and tabulate the number of times that the nearest point is a case. For maps where cases are randomly located with respect to controls, the expected number of such case–case pairs is 20(19/69) = 5.5. To see why this is, recognize that when examining a particular case, there are 19 + 50 = 69 other points on the map, and 19 of them are cases. Thus the probability that a point chosen randomly from these other points is a case is equal to 19/69. Since we repeat this 20 times (once for each case), the expected number of case–case pairs is 20(19/69) = 5.5.

Observing anything more than 5.5 case–case pairs indicates a tendency for cases to cluster near one another. If the number of case–case pairs is less than 5.5, this indicates a tendency for cases to be closer to controls than would be expected by chance alone. There is a tendency for individuals to choose case locations that are not near other case locations; hence the number of case–case pairs on maps created by individuals is often less than that expected in a random pattern. This underscores the ineffectiveness of people at assessing spatial randomness, and helps to motivate the need for spatial statistical methods.

In this particular example, there is another way to assess the spatial pattern of cases. Since the population (and the spatial distribution of controls) is random, the location of cases should also be random. There are 25 cells on the map. For each cell, count the number of cases in the cell, and give the cell a score according to those in Table 1.1.

Table 1.1   Cell scores (see Section 5.1 for a more detailed explanation of how scores were generated)

Number of points in a cell	Score
0	0.8
1	0.05
2	1.8
3	6.05
4	12.8

The total score for your map is simply the sum of the cell scores. A computer locating 20 cases at random would produce an average map score of 24. Case distributions that are more clustered than random will have scores greater than 24, and when the pattern of cases is more spread out than random, the score will be less than 24. In general (but not always!) people tend to have difficulty producing a random map of cases, and they spread out their cases too much, consequently ending up with map scores that are less than 24.

1.3.2 Distinguishing Random Coin Tosses from a Series Constructed to Appear Random

A now-classic experiment (a form of which is described by Gelman and Nolan, 2017) designed to illustrate the nature of randomness involves two students in a classroom. They are each asked to toss a coin 100 times, and record the sequence of heads and tails. One student does so; the other decides that this is too much work, and makes up a sequence of 100 H's and T's, representing heads and tails, respectively. Can you tell from the sequences in Figure 1.4 which sequence is which?

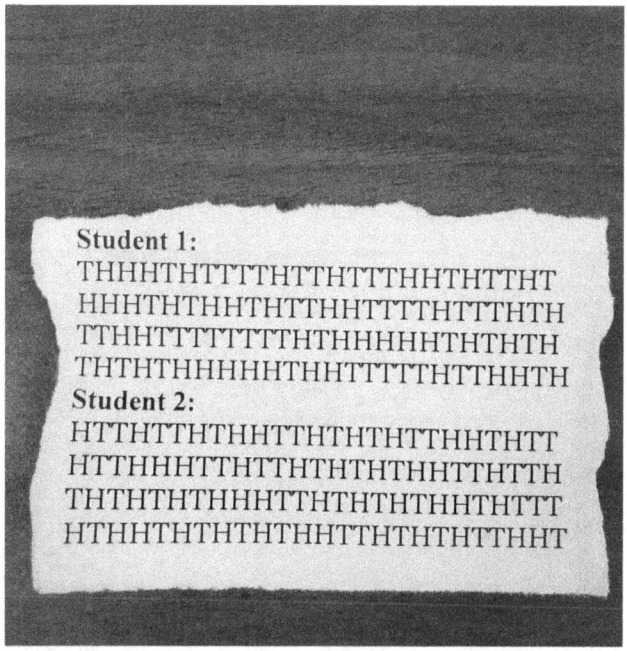

Figure 1.4 The 100 coin tosses for two students

Source: Gelman and Nolan (2017). Reproduced with permission of Oxford University Press.

When asked to create a sequence of random tosses, individuals typically do not put in runs of H's or T's that are long enough. Instead, they tend to alternate between "T" and "H" too much. A count of the number of runs in the sequence therefore allows one to tell which sequence is which. A random sequence of n tosses will have an average of $n/2$ runs. There are 56 runs in Student 1's list. There are 74 runs in Student 2's list. This leads us to believe that it is Student 2 who has contrived the sequence. A more formal assessment uses the runs test, which makes use of the fact that for random sequences, the expected number of runs is $n/2$, and the standard deviation of the number of runs is $\sqrt{n}/2$. Therefore we can calculate z-scores for each student. For Student 1,

$z = (56 - 50)/(10/2) = 6/5 = 1.2$. For Student 2, $z = (74 - 50)/(10/2) = 24/5 = 4.8$. The first sequence is consistent with a randomly generated sequence; the observed value of z does not exceed the critical value (of, say, 1.96 in a two-tailed test with $\alpha = 0.05$). The second student had many more runs than could be expected by chance alone; the z-score of 4.8 convincingly rejects the null hypothesis of randomness.

1.3.3 Distinguishing a Random Pattern of Points from a Pattern of Glowworm Locations

There are two patterns in Figure 1.5; these are cited in Pinker (2011), who in turn was recounting a discussion in Gould (1992). One of them has been produced by a computer, using a random number generator to produce a random pattern. The other is a modeled set of locations (as opposed to an *actual* set of locations) of glowworms on the ceiling of Waitomo Cave, in New Zealand. Can you tell which is which?

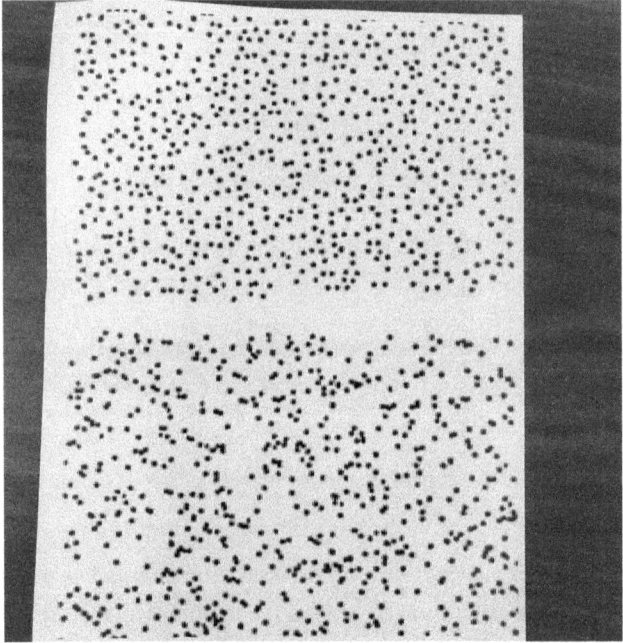

Figure 1.5   Glowworm locations and random locations

Source: Gould (1992)

The map at the top of the figure is the map of the presumed/modeled spatial distribution of glowworms; the map at the bottom is a completely random, computer-generated pattern. The glowworms spread out as they compete for food, establishing their territories.

It is easy to think that this is the random pattern – after all, the other map seems to have clusters, voids, filaments, and other features. But those features are actually characteristics of spatial randomness – they appear by chance alone. Again, the eye is good at organizing information and seeing patterns – even when locations have been decided independently! Of course in some sense patterns *do* exist in randomly generated data – there *are* clusters, since some points are very close to one another. There *are* filaments, since some points align along string-like paths. But these have occurred by chance. It would actually be very unusual for points placed at random to result in a map like the glowworm map.

This raises the question of what we really mean by randomness. A point pattern in space that is random has two characteristics: (1) every small subarea of a given size has an equal probability of receiving a point (a so-called "first-order" property), and (2) every point has a location that is independent of the location of every other point (a "second-order" property).

1.3.4 Finding the Mean for Spatial Data

In the previous examples, we have explored perceptions of randomness. Here we consider how complex even the simplest task of finding the mean for a set of spatially dependent data can be. Consider the following set of three sample observations, taken at different geographic locations within a county:

$$x_1 = 0.9889; x_2 = 1.388; x_3 = -0.6242 \tag{1.1}$$

These might, for example, represent annual precipitation totals in inches or centimeters for the three locations, represented as deviations in inches from the previous year's average precipitation value for that county. Further imagine that locations 1 and 2 are nearby to one another, and that they are both much farther away from location 3.

Our modest goal is to estimate the true and unknown population mean of the deviations, μ. That is, we want to decide whether precipitation has, overall, gone up ($\mu > 0$), down ($\mu < 0$), or remained the same ($\mu = 0$). An immediate reaction is to find the simple mean, or average, of the three observations: $\bar{x} = (0.9889 + 1.388 - 0.6242)/3 = 0.5843$. This approach ignores an important aspect of geographic data – observations that are near to one another in space are often correlated with one another.

Note that observations 1 and 2 have values that are not dissimilar from one another, and those locations are also near to one another. In principle we might want to down-weight these two observations – there is a redundancy in the information they convey. If all three observations were independent, we would give each an identical weight of 1/3 in computing the sample mean. Here, however, we would like the weights for observations 1 and 2 to be less than 1/3, and the weight attached to observation 3 to be more than 1/3.

In addition to estimating the mean, we may also want to make some statement about our level of confidence in the sample mean. This is usually evaluated by means of (a) the standard normal, z-distribution, when the true standard deviation, σ, is known, or (b) a t-distribution with $n-1$ degrees of freedom, in the much more common case where the true standard deviation is unknown (if we didn't know the true mean, it would certainly be unusual to know the true standard deviation!).

For this example, the corresponding 95% confidence intervals would be, for cases (a) and (b), respectively:

(a) $\bar{x} \pm z\sigma / \sqrt{n} = 0.5843 \pm 1.96 \quad (1/\sqrt{3}) = 0.5843 \pm 1.132$

$$(1.2)$$

(b) $\bar{x} \pm ts / \sqrt{n} = 0.5843 \pm 4.302 \quad (1.065 / \sqrt{3}) = 0.5843 \pm 2.645$

where s is the sample deviation, 1.96 is the critical value of z that leaves 2.5% of the area in each tail of the distribution, and $t = 4.302$ is the corresponding value (leaving 2.5% of the area in each tail) for the t-distribution with $n-1 = 2$ degrees of freedom. With just three observations here we, not surprisingly, have wide confidence intervals.

Not only should we consider down weighting observations 1 and 2 in their contributions to the sample mean, but we should also recognize that the confidence intervals just derived are too narrow – effectively, we don't really have three independent observations. Because of the geographic closeness and correlation between observations 1 and 2, we have the equivalent of a sample size with less than three independent observations. Consequently, the actual confidence intervals will be wider than those above.

Finally, if we were to test the null hypothesis that the true mean was equal to zero (against the two-tailed alternative that μ is not equal to zero) in the usual way, we would either use (a) a z-test, with the observed z-statistic:

$$z = 0.5843 / \left(1/\sqrt{3}\right) = 1.012 \qquad\qquad (1.3)$$

or (b) a t-test, with the observed t-statistic:

$$t = 0.5843 / \left(1.065 / \sqrt{3}\right) = 0.95 \qquad\qquad (1.4)$$

Since these observed statistics are less than their critical values (of 1.96, and 4.302 for the z- and t-tests, respectively), we fail to reject the null hypothesis. The p-value associated with the two-tailed t-test is 0.221 (2) = 0.442, and that associated with the z-test is 0.3115; our observations are not particularly unusual, if the null hypothesis is true. These are both clearly greater than 0.05 and consistent with failing to reject the null hypothesis. However, the z- and t-statistics are higher, and the p-values are lower than they would be if spatial dependence were taken into account. In Chapter 8 we will learn how to improve these estimates by taking into account the effects of spatial dependence.

2

A Quick Review of Some Key Material from Introductory Statistics

Chapter overview

In this chapter, you will learn about:

- Probability distributions
- Distribution of sample means
- Confidence intervals
- Hypothesis testing

2.1 Introduction

We will first have a quick refresher of some of the highlights of an elementary first course in statistics. Introductory courses in statistics are often divided into two major parts – a relatively short section on *descriptive* statistics, and a longer section on *inferential* statistics. The intention of the latter is to convey the basic concepts and methods associated with the goal of making inferences about complete populations from sample data. Descriptive statistics include numerical measures that summarize a dataset. These include the sample mean, $\bar{x} = n^{-1} \sum x_i$, and the sample variance, $s^2 = (n-1)^{-1} \sum (x_i - \bar{x})^2$ (where the standard deviation, s, is simply the square root of the variance). The mean and variance are measures of the central tendency and dispersion, respectively, for data distributions.

These measures may be combined in various ways. The coefficient of variation ($s/\bar{x}$) is a unitless measure of relative dispersion. A standardized score, or z-score, is a unitless measure of how many standard deviations an observation lies away from the sample mean.

For example, with the five observations $x_1 = 3$, $x_2 = 6$, $x_3 = 10$, $x_4 = 5$, and $x_5 = 6$, we find that the sample mean is equal to 6, and the variance is equal to $26/4 = 6.5$. The first observation has a z-score of $(x_1 - \bar{x})/s = (3 - 6)/\sqrt{6.5} = -1.18$; it lies more than one standard deviation below the mean of 6.

There are geographical analogues of these measures of central tendency and dispersion, and these are discussed further in the next chapter.

There are of course many more descriptive statistics, and we will encounter some of them along the way.

2.2 Probability and Probability Distributions

Probability distributions may be thought of as theoretical histograms. Certain forms of such histograms arise repeatedly and frequently in practice because they have common features that follow from common assumptions that represent various real-world processes well.

2.2.1 Binomial Distribution

Binomial processes are characterized by the following assumptions: (1) the experiment consists of n trials, (2) the trials are independent, (3) each trial is characterized by two possible outcomes (e.g. heads or tails, rain or no rain, success or failure, etc.), and (4) the probability of success on each trial (p) is constant. With these characteristics, the probability of x successes in n trials is

$$\Pr(X = x) = \binom{n}{x} p^x (1-p)^{n-x} \; ; x = 0,1,\ldots,n \tag{2.1}$$

This arises from the fact that the chance of getting x successes and $n - x$ failures is $p^x(1-p)^{n-x}$, and the number of ways in which this could happen (i.e., the number of ways that x successes and $n - x$ failures could be rearranged) is $\binom{n}{x} = n!/[x!(n - x)!]$.

This probability distribution is discrete; the values that the random variable X can take consist of a finite set of possibilities (0, 1, ... , n). The equation describing the theoretical frequency distribution (or, if you prefer, a theoretical histogram) for a discrete random variable is sometimes referred to as a *probability mass function*.

The theoretical mean of the binomial distribution is $\mu = np$, and the variance is equal to $\sigma^2 = np(1 - p)$. These theoretical means and variances characterize distributions that follow all of the assumptions governing their generation precisely. If very large samples were generated with these assumptions, the sample means and variances would approach their theoretical values. Greek letters are used to denote the theoretical mean and variances, while sample means and variances are denoted by $\bar{x}$ and s^2, respectively.

2.2.2 Poisson Distribution

The Poisson distribution is also a discrete distribution; its uses include the modeling of rare events, and the modeling of events that are random in time or space. It is also used as an approximation to the binomial distribution when n is large, p is small, and the expected number of events is small (generally less than about 5).
   Its probability mass function is

$$\Pr(X = x) = \frac{e^{-\lambda}\lambda^x}{x!} ; x = 0,1,2... \tag{2.2}$$

where e is the constant 2.718... and λ is a parameter equal to the expected number of events. The theoretical mean and variance are both equal to λ.

Example

The average number of accidents at an intersection each month is equal to 3.2. Find the probability that there are no accidents next month at the intersection.

Solution

$\lambda = 3.2$. The answer is equal to

$$\frac{e^{-3.2}3.2^0}{0!} = 0.0408 \tag{2.3}$$

This can be generalized for time periods with other lengths. For example, the probability that there is exactly one accident in the next 15 days is, assuming a month to be equal to 30 days,

$$\frac{e^{-3.2\left(\frac{15}{30}\right)}\left(3.2 \times \frac{15}{30}\right)^1}{1!} = 0.323 \tag{2.4}$$

where $3.2(15/30) = 1.6$ is the number of accidents expected over the 15-day period.
   The Poisson distribution also describes random processes in space. If there are, on average, three cases of disease every 20 square miles, and if cases are distributed randomly, we

can use the Poisson distribution to tell us how unusual it would be to observe four or more cases in a ten square mile area. The first step is to ask how many cases we would *expect* in this ten square mile area. The answer is 1.5 (and this could be found formally by solving $3/20 = \lambda/10$). Then we have

$$1 - \sum_{x=0}^{3} \frac{e^{-1.5}1.5^{x}}{x!} = 0.0656 \tag{2.5}$$

When using the Poisson distribution, often the first step is to determine what to use for the parameter λ, the expected number of events. If you know that there are 2.1 floods per decade, and are interested in the likelihood of more than one flood in the next four years, the first step is to determine the *expected number* of floods during the next four years, our time period of interest. Thus

$$\frac{2.1}{10} = \frac{\lambda}{4} \tag{2.6}$$

leads to $\lambda = 0.84$. The same is true when working with the Poisson distribution for spatial problems. Suppose that we know there are three cases of a disease in a 20 square mile region, and we want to know the likelihood of three or more cases in our town, which occupies 15 square miles. The appropriate value of λ comes from

$$\frac{3}{20} = \frac{\lambda}{15} \tag{2.7}$$

which implies that we would expect 2.25 cases in our town.

2.2.3 Normal Distribution

Continuous distributions take on an infinite number of possible values across a range. One way to think of them is as histograms with an infinite number of infinitely narrow vertical bars. They are specified in such a way that areas under the curves represent probabilities; the total area under a continuous probability curve is equal to 1.

The normal distribution is the most important example of a continuous probability distribution. It takes on values between minus infinity and infinity; the equation describing its form (equations for continuous probability distributions are denoted with the notation $f(x)$ and are referred to as *probability density functions*, or *pdfs*) is

$$f(x) = \frac{1}{\sqrt{2\pi}\sigma} e^{-\frac{(x-\mu)^2}{2\sigma^2}} \tag{2.8}$$

where μ and σ^2 describe the shape of the curve and represent the mean and variance, respectively. This curve traces out the familiar symmetric, bell-shaped curve also known as the Gaussian distribution. This curve simultaneously traces out how the likelihood of specific values changes with x.

The probability of obtaining a value of x between say a and b is equal to the area under this function that lies between a and b. In calculus terms the area under the curve is an integral (for more details, see Appendix A), and, in turn, an integral is nothing more than an infinite sum – we are summing the areas of that infinite number of very thin vertical bars. Thus

$$\Pr(a \le x \le b) = \int_a^b \frac{1}{\sqrt{2\pi\sigma^2}} e^{-\frac{(x-\mu)^2}{2\sigma^2}} dx \tag{2.9}$$

It is not possible to solve for this integral analytically and instead we resort to tables of the *standard normal distribution*, which has a mean of 0 and a standard deviation (and variance) of 1.

The first step in finding probabilities for the normal distribution is to convert the normal distribution in question into the standard normal distribution, and this is done using z-scores. For example, suppose that daily mean temperatures during the month of July in a location are normally distributed, with mean 72 degrees Fahrenheit and standard deviation 8 degrees. If we are interested in the probability that the mean temperature on a randomly chosen day in July is between 76 and 84, we can proceed as follows. We know that the answer is equal to the area between $a = 76$ and $b = 84$ that lies under the symmetric, bell-shaped curve centered at $\mu = 72$, with $\sigma = 8$. We first convert this into a question involving z-scores. The value 76 is $z = 0.5$ standard deviations above the mean, and 84 is $z = 1.5$ standard deviations above the mean. The desired area between temperatures of 76 and 84 is the same as the area between $z = 0.5$ and $z = 1.5$ on the standard normal curve. The reader should consult a table (or online applet) to verify that this probability is equal to 0.2417. Readers should also be cautioned that some tables give *tail areas* (i.e., $\Pr(Z > z)$), while other tables are arranged differently, and give the probability that Z is between 0 and z (i.e., $\Pr(0 < Z < z)$).

Questions involving percentiles are also easily answered. The 60th percentile of a normally distributed variable may be found by first finding the z-score that leaves 40% of the area in the right tail of the distribution; in this case, the z-score is $z = +0.253$. For the example above we know the 60th percentile is that temperature which is 0.253 standard deviations above the mean. Thus $72 + 0.253(8) = 74.024$ degrees is the 60th percentile.

2.2.4 Exponential Distribution

Many distributions in the social and physical sciences have a strong positive skewness – there are a large number of small values, and a small number of large values. A common example is income – there are a small number of people making a lot of money. In geography, distances – distances to go shopping, to commute, to move a residence – are strongly skewed. One simple distribution that captures these tendencies is the exponential distribution. It is a one-parameter distribution:

$$f(x) = \lambda e^{-\lambda x}; \ x \geq 0 \tag{2.10}$$

with the parameter λ indicating the steepness of the decline of the probability density with increasing x (with higher values of λ associated with steeper rates of decline). The expected value of x (denoted E[x]) is equal to the reciprocal of the parameter: $\mu = $ E[x] $= 1/\lambda$. A more detailed discussion of parameter estimation may be found in Appendix A. To fit an exponential distribution to a set of data the parameter λ may be estimated as the reciprocal of the sample mean: $\hat{\lambda} = 1/\bar{x}$ (where the caret ("^") indicates an estimate of the unknown value).

To find the areas under exponential distributions that are associated with probabilities, we make use of the *cumulative distribution function* (cdf):

$$\Pr(X < x) = F(x) = 1 - e^{-\lambda x} \tag{2.11}$$

where the general notation $F(x)$ is used for cdfs. Note that $F(x) = 0$ when $x = 0$, and $F(x)$ approaches 1 as x gets large. To find the probability that a value from the exponential distribution lies between a and b, recognize that this is equal to the probability that x is less than b, *minus* the probability that x is less than a:

$$\Pr(a \leq x \leq b) = F(b) - F(a) = \left(1 - e^{-\lambda b}\right) - \left(1 - e^{-\lambda a}\right) = e^{-\lambda a} - e^{-\lambda b} \tag{2.12}$$

The probability that an exponential variate is *greater* than x is simply equal to $e^{-\lambda x}$. The probability of obtaining a value between a and b can also be thought of as the probability of getting a value greater than a, *minus* the probability of obtaining a value greater than b:

$$\Pr(a \leq x \leq b) = e^{-\lambda a} - e^{-\lambda b} \tag{2.13}$$

For example, if $\lambda = 1/4$, the expected value or theoretical mean of the variable is equal to 4. The probability of a value lying in the range (3, 6) is equal to $e^{-3(1/4)} - e^{-6(1/4)} = 0.2492$.

The pth percentile (where p is expressed as a percentage between 0 and 100) of the exponential distribution may be obtained by setting the cumulative distribution function equal to $p/100$ and then solving for x:

$$F(x) = \Pr(X < x) = 1 - e^{-\lambda x} = \frac{p}{100} \tag{2.14}$$

This implies that $1 - (p/100) = e^{-\lambda x}$. Taking the natural logs of both sides, $\ln(1 - p/100) = -\lambda x$, and $x = -\ln(1 - p/100)/\lambda$. For example, the 35th percentile of an exponential distribution with parameter $\lambda = 1/4$ is equal to $-\ln(1 - 35/100)/(1/4) = -4 \ln(0.65) = 1.723$.

The exponential distribution has what is known as the *memoryless* property. If for example the time between residential moves follows an exponential distribution, then the probability that a person will move in the next week is *independent* of how long they have been there already. When the next occurrence will happen does not depend upon how long it has been since the last occurrence.

There is also a connection between exponential and Poisson distributions. In particular, people arriving at a location randomly in time can be counted using a Poisson distribution. The distribution of interarrival times (i.e., the length of time elapsed between the random arrival of successive people) follows an exponential distribution.

2.3 The Distribution of Sample Means

The central limit theorem states that the sum of independent variables will tend toward a normal distribution as the number of observations gets large. This is true no matter what the distribution of the underlying variables is. A fair die for example has a (discrete) uniform distribution – six equally likely outcomes. But the histogram associated with the sum of the outcomes for ten rolls of the die will look a lot like the normal distribution – ten would be an unlikely outcome (all ones), as would sixty (all sixes), and the sum would frequently be in the neighborhood of 35 (the average of 1 through 6 is equal to 3.5, which is then multiplied by 10).

Perhaps the most important implication of the central limit theorem is that sample means have a normal distribution (if the sample is sufficiently large, where "large" is usually taken to mean about 30 or more). In particular, sample means have normal distributions with mean equal to the true mean μ, and variance equal to the variance of the underlying variable, divided by the sample size, n:

$$\sigma_{\bar{x}}^2 = \frac{\sigma^2}{n} \tag{2.15}$$

Note that there are three instances of the word "mean" in the previous sentence; each has a different meaning. In addition to the sample mean, there is a "mean of sample means" – this second instance of the word refers to the fact that if there were a collection of sample means, this collection would itself have a mean. Finally, the third instance refers to the fact that there is some true underlying mean associated with the enumeration of all individuals in the total population.

From Equation 2.15 we can see that the variance of the distribution of the sample means will be high when (a) the variance of the observations (σ^2) is itself high, (b) the sample size n is relatively low, or (c) a mixture of both (a) and (b).

2.4 Confidence Intervals

A significant consequence of the fact that sample means have normal distributions (with known or estimable variance) is that when we find a sample mean ($\bar{x}$) we not only have an estimate of the mean (μ) of the complete population (which corresponds to the set of all individuals who could possibly be sampled), but also have a calculable measure of our degree of confidence in that estimate. Sample means are taken from a normal distribution, with mean μ and variance given by Equation 2.15. Thus

$$\Pr\left(\mu - \frac{z\sigma}{\sqrt{n}} \leq \bar{x} \leq \mu + \frac{z\sigma}{\sqrt{n}}\right) = 1 - 2\Phi(-z) \tag{2.16}$$

where $\Phi(-z)$ is the tail area of the standard normal distribution to the left of $-z$. This can be rearranged as

$$\Pr\left(\bar{x} - \frac{z\sigma}{\sqrt{n}} \leq \mu \leq \bar{x} + \frac{z\sigma}{\sqrt{n}}\right) = 1 - 2\Phi(-z) \tag{2.17}$$

For example, the 95% confidence interval for the sample mean is found using $z = 1.96$:

$$\Pr\left(\bar{x} - \frac{1.96\sigma}{\sqrt{n}} \leq \mu \leq \bar{x} + \frac{1.96\sigma}{\sqrt{n}}\right) = 0.95 \tag{2.18}$$

As a practical matter, the standard deviation is not usually known, and σ is replaced with the sample standard deviation, s. We can interpret a 95% confidence interval for the sample mean as follows: 95% of the time, confidence intervals constructed in this manner will contain the true mean; 5% of the time we will obtain an unusual sample, and the true mean will not be contained within the interval.

───────────────────（ **Example** ）───────────────────

A sample of 36 people reveals that their migration distances had a mean of 50 miles, and a standard deviation of 30 miles. Find a 90% confidence interval for the mean.

Solution

For a 90% confidence interval, we use $z = \pm1.645$, since that value leaves 5% of the area in each tail. The 90% confidence interval is 50 miles $\pm$ 1.645(30/$\sqrt{36}$) = 50 $\pm$ 8.225, or (41.775, 58.225). We dont know the true mean but there is a 90% chance that the true mean is contained within this interval.

When the sample size is small (less than about 30) the sums (and average) have not yet converged closely to the normal distribution; the t-distribution is used in its place, with $n - 1$ degrees of freedom. Distributions can be similarly constructed for sample proportions, for the difference in two sample means, and for the difference in two sample proportions.

───

2.5 Hypothesis Testing

Tests of hypotheses about means can also be carried out on the basis of the fact that sample means have normal distributions. Hypothesis testing begins by stating a null hypothesis (H_0) – this is often a hypothesis that we set out to reject (akin to a straw man we hope to, using the evidence, knock down).

The second step is to state an alternative hypothesis (usually notated as H_1 or H_a) – this can be either one-sided or two-sided, depending upon whether we have some a priori idea that the mean will be less than, greater than, or different from the value stated in the null hypothesis. With the null hypothesis H_0: $\mu = \mu_0$, a two-sided alternative hypothesis would be employed if we have no a priori idea of the direction the mean will deviate from the null value. In that case, H_1: $\mu \neq \mu_0$. If we think that the mean will turn out to be greater (or less) than μ_0, we use the one-sided alternative, H_1: $\mu > \mu_0$ (or H_1: $\mu < \mu_0$).

The next step is to choose α, the probability of making a Type I error, where a true null hypothesis is rejected. We have control over this type of error, and the most common choice is $\alpha = 0.05$; other common choices are 0.01 and 0.1. It should also be kept in mind that the lower the value chosen for α, the higher will be β, the probability of a Type II error (where a false null hypothesis is accepted). If α is very low, we wont make the mistake of rejecting true null hypotheses, but we also wont reject null hypotheses when they are false! The *power* of a statistical test is equal to $1 - \beta$; it is the probability that we reject a false null hypothesis, and wed like this power to be high. We will return to the discussion of statistical power when we compare alternative tests designed for the detection of statistical clustering.

Next, an observed test statistic is chosen and then compared with the critical value (in the case of a one-sided alternative) or critical values (for two-sided alternatives). The test statistic for large, one-sample hypotheses about the mean is the z-test, $z = (\bar{x} - \mu_0)/(s/\sqrt{n})$. This statistic derives directly from the fact that means have normal distributions with mean μ and standard deviation $\sigma/\sqrt{n}$. The test statistic tells us how far our observed mean is from the hypothesized mean (μ_0), in terms of the number of standard deviations (of the distribution of sample means).

The critical values establish a rejection area (or areas) in the tail or tails of the distribution. They constitute the boundary between rejecting and failing to reject the null hypothesis. A decision can then be made; if the observed statistic falls into the tail area as determined by the critical value, the null hypothesis is rejected; otherwise we fail to reject it. For our z-test involving a hypothesis about the true population mean, the critical value is that value (or values, in the case of two-sided tests) of z that leaves a total area of α in the tail (or tails, in the case of two-sided tests) of the standard normal distribution.

The final step is to find the p-value, which is the likelihood of observing a test statistic more extreme or more unusual than the one actually observed, if the null hypothesis is true. This is found by simply using the observed test statistic to find the area in the tail that is associated with it (and then the result is multiplied by 2, if the alternative hypothesis is two-sided). This gives us a little added information. Instead of simply rejecting a null hypothesis, or failing to reject it, we now have a numerical measure giving us an idea of how unusual our result is if the null hypothesis is true. The p-value is a numerical measure of how convincingly we reject the null hypothesis (if we do), or how close we come to rejecting it (if we dont reject it).

Of course hypothesis tests are also constructed for cases with two samples (where the interest is often in the difference in means), for cases with small samples (in which case a t-distribution is used), and for many other situations (including, for example, tests involving proportions).

-------------------------(**Example**)-------------------------

The mean number of children residing in households in a county is equal to 1.8. A researcher believes that, in her town, families may be larger than the countywide average. Data are collected for 42 households; the mean number of children is equal to 2.25, and the sample standard deviation equals 2.0. Test the null hypothesis that the true mean is equal to 1.8. Use $\alpha = 0.05$.

Solution

$H_0: \mu = 1.8$. $H_1: \mu > 1.8$. (This is a one-sided alternative, since there is an a priori hypothesis that the mean will be greater than the countywide average.) We will use a Type I error probability of $\alpha = 0.05$, and our test statistic, based upon the observed data, is $z = (2.25 - 1.8)/(2/\sqrt{42})$ $= 1.458$. The critical value of z is equal to 1.645; this z-value leaves 5% of the area in the tail, to the right. We fail to reject the null hypothesis, since the observed test statistic is less than the critical value. The p-value is equal to 0.0724; this is the area to the right of 1.458 in the standard normal distribution; 7.24% of the time, we would expect a result more extreme than the one we observed, if the households in her town were not larger than the countywide average. This is not unusual enough to reject the null hypothesis. Only when $p < \alpha$ is the null hypothesis rejected.

There is a correspondence between confidence intervals and two-sided alternative hypotheses. Suppose that we test the null hypotheses that a mean migration distance when moving is equal to 10 km, against the alternative hypothesis that this is not true. We collect data from 49 individuals, and find a sample mean of 12 km, and a sample standard deviation of 7 km. Hypothesis testing would yield a test statistic of $(12-10)/(7/\sqrt{49}) = 2.00$ and for a two-sided test with $\alpha = 0.05$, we would reject the null hypothesis. The p-value would be equal to $(2) \times 0.0228 = 0.0456$. If we construct a 95% confidence interval for the sample mean, we have $12 \pm 1.96(7/\sqrt{49}) = 12 \pm 1.96 = (10.04, 13.96)$. Note that this interval does *not* include the hypothesized mean (10). Rejecting a null hypothesis is consistent with finding confidence intervals that do not contain the hypothesized mean.

Exercises

1. Find the 90% confidence interval for the sample mean, when $n = 40$, $\bar{x} = 34.2$, and $\sigma = 12.1$.

2. Fifty soil samples are taken; the mean pH level is 7.23, and the standard deviation is 0.72. Find a 95% confidence interval for the sample mean. Also test the null hypothesis that the true pH level is equal to 7.0 and find the p-value. Use $\alpha = 0.05$.

3. For the following data: 3, 6, 11, 13, 17, 5, 2, 1, 10, 18, find the mean, standard deviation, variance, and coefficient of variation. Find z-scores for the first three observations.

4. The probability of a household moving each year is equal to 0.135. What is the probability that exactly three households move during a year, out of a total of ten that are surveyed? What is the mean number of households you would *expect* to move next year? What is the probability that less than two households move next year, out of a total of eight surveyed households?

5. The expected number of tornados each decade in a town in the Midwestern United States is 1.2. Find (a) the probability of no tornadoes in the next six years, and (b) the probability of fewer than three tornados in the next nine years. Hint: To use the Poisson distribution, the first step is to find λ, the expected number of events during the time period of interest.

6. The traffic coming through an intersection during a particular hour of the day is normally distributed, with mean equal to 38 and standard deviation equal to 10. Find the probability that the number of cars coming through the intersection is (a) between 33 and 50, (b) less than 29. Also find the 70th percentile of the distribution.

7. The distance traveled by consumers to a local supermarket is exponentially distributed, with mean 4.3 miles. Find the probability that a consumer travels (a) more than 6 miles, (b) less than 3 miles, and (c) between 4 and 7 miles. Also find the 80th percentile of this distribution. What is the meaning of this percentile? Hint: To find the 80th percentile, recognize that you are trying to find the value (x) such that 80% of all observations are less than it. You know the area under the curve (0.8), and need to use that, along with λ, to find the unknown value of x.

Further reading

There are many introductory treatments of statistics, and a large number of good accounts of introductory statistics for geographers. The following is simply a short list for geographers that is not by any means comprehensive, and there is no intent to deliberately omit the authors of the many other fine books!

Burt, J.E., Barber, G.M., and Rigby, D.L. (2020) *Elementary Statistics for Geographers* (3rd edition). New York: Guilford Press.

McGrew, Jr., J.C., Lembo, Jr., A.J., and Monroe, C.B. (2014) *An Introduction to Statistical Problem Solving in Geography* (3rd edition). Long Grove, IL: Waveland Press.

Rogerson, P. (2020) *Statistical Methods for Geography: A Students Guide* (5th edition). London: Sage.

Online resources

Visit **https://study.sagepub.com/rogersonspatialstatistics** for resources which supplement the material presented in this chapter. Students can find datasets and additional exercises, which can be used to practice the techniques covered in this chapter, as well as a chapter introduction video.

Lecturers and instructors can find exercises and conceptual prompts to facilitate classroom discussion and practice.

3

Some Selected Measures for Descriptive Spatial Statistical Analysis

Chapter overview

In this chapter, you will learn about:

- Centers of population
- Measures of dispersion
- Measures of "coastiality"
- Geographic centers
- Measures of inequality

The center is a fundamental geographic concept; just as the mean is one of the most basic ways to describe a set of numerical data, the center of a set of geographic locations is a basic way to describe those locations.

Spatial measures of centrality are associated with measures of accessibility and can therefore be useful in solving location problems. What central location, for example, minimizes the sum of distances people have to travel to get there?

Here we will discuss the two most fundamental types of spatial center – centers of population, and geographic centers. For each of these types, we will see that the relatively simple solutions for problems in two dimensions become somewhat involved when we consider the center of locations on the surface of a three-dimensional sphere.

We begin with a discussion of centers of population. After a brief review of centers for two-dimensional representations of space, we focus on finding centers for large regions of space (e.g., countries), where a three-dimensional solution is called for. Here we define different types of centers, provide methods of calculation, and provide results for some examples. Locations of centers may be important in their own right, in providing locations with maximum accessibility, but they also can provide a good idea of how geographical distributions are changing over time. They can also provide good insights into the spatial distribution of different groups of people (e.g., particular age groups, ethnicities, etc.).

3.1 Centers of Population

For any region, the center of population is the location at which a map of the locations of all individuals would balance, if equal weights were assigned to each individual location. It is therefore also known as the *center of gravity*. If you were to hold up a physical representation of all locations that had weights that were proportional to population, the mean center would be the point of balance.

The center of gravity is often known as a *mean center*, and has the property that the sum of squared distances that people have to travel to get to the center is minimized (just as the average or mean of a set of data is that location on the number line that minimizes the sum of squared distances along the number line from all the values in the dataset to the mean).

3.1.1 Mean Center

Mean centers of population are found by weighting locations by their population. Typically all of the people in a subregion are assigned to a particular location within that subregion – that could be the geographic center of the subregion, or, more commonly, it could be the location of the most populous place within that subregion. Clearly, the smaller the subregions, and the higher the geographic resolution, the more accurate the center's location will be.

The mean center $(\bar{x}, \bar{y})$ is found simply as an average of the x- and y-coordinates of all locations. In most applications the coordinates of locations are weighted by population, and in that case a weighted average of the x- and y-coordinates is taken. For example, the center of population for a county may be found by weighting the x- and y-coordinates of the centers of the townships that make up the county by the populations of the townships. This makes the assumption that all people in the township live at the center of that township. Thus

$$\bar{x} = \frac{\sum_{i=1}^{n} w_i x_i}{\sum_{i=1}^{n} w_i}; \; \bar{y} = \frac{\sum_{i=1}^{n} w_i y_i}{\sum_{i=1}^{n} w_i} \tag{3.1}$$

where location i has coordinates (x_i, y_i) and population w_i.

3.1.2 Median Center

The median center achieves the more intuitive objective of minimizing aggregate travel, or sum of distances (as opposed to the mean center, which minimizes the sum of squared distances). The solution is also known in the field of location theory as the solution to the p-median problem, for the special case where $p = 1$ (the p-median problem is the problem of locating p facilities to minimize the distance traveled to the facilities, assuming that people go to the nearest facility).

However, satisfying the attractive objective comes at the cost of a more complex calculation to determine the location. Weiszfeld (1937) provided an iterative numerical solution for finding the point of minimum aggregate travel. The median center is found by beginning with a guess which we denote as $(x^{(0)}, y^{(0)})$. Since the mean center is easy to calculate, it is often used as the initial guess. Then new x- and y-coordinates are found from

$$x^{(1)} = \frac{\sum_{i=1}^{n} \frac{w_i x_i}{d_i}}{\sum_{i=1}^{n} \frac{w_i}{d_i}} \; ; \; y^{(1)} = \frac{\sum_{i=1}^{n} \frac{w_i y_i}{d_i}}{\sum_{i=1}^{n} \frac{w_i}{d_i}} \tag{3.2}$$

where d_i is the Euclidean distance from location i to the current estimate of the median center. This result is used as an updated estimate of the median center, and a new estimate $x^{(2)}, y^{(2)}$ is found from the same equation (using the same weights, and the same x_i, y_i locations, but the newly calculated distances from the locations to the newly estimated location for the median center). After several iterations, the change in the location for the median center will be very small. The iterative process is terminated when the further improvement in the estimate of the median's location is negligible.

Note that the locations are weighted by the quantity (w_i/d_i) which has, in the case of population, been interpreted as a measure of a location's *population potential* (see, for example, the social physics work of Stewart and Warntz, 1958).

Of course there are different ways to measure distance. With x- and y-coordinates, one straightforward way to measure distance is to use the definition of Euclidean distance; the distance between two points i and j may be thought of as the hypotenuse of a right triangle, and hence

$$d_{ij} = \sqrt{\left(x_i - x_j\right)^2 + \left(y_i - y_j\right)^2} \tag{3.3}$$

For cities with grid-like street patterns, a Manhattan distance may be more appropriate:

$$d_{ij} = \left|x_i - x_j\right| + \left|y_i - y_j\right| \tag{3.4}$$

These descriptive measures of central tendency for spatial data assume a relatively small spatial scale (small enough that the earth can be assumed flat over the range of the study area). When the data cover a larger scale, the objectives will be to minimize the sum of great circle distances, or the sum of squared great circle distances (see, e.g., Plane and Rogerson, 2015; Rogerson, 2015a).

It should also be noted that the US Bureau of the Census has used an alternative definition of the median center – namely, the point that serves as the intersection of the 50th percentile (i.e., the median) of observations based on a longitudinal axis (i.e., the point that has half of the population to the east, and half to the west), and the 50th percentile of observations based on a latitudinal axis (so that half of the population lies to the north, and half to the south). See, for example, the description and historical results provided by the US Bureau of the Census (www2.census.gov/geo/pdfs/reference/cenpop2010/COP2010_documentation.pdf).

The concept of center of population and its variants is useful in tracking demographic change. Plane and Rogerson (2015) investigated how the individual components of population change (births, deaths, international migration, and internal migration) have affected the movement of the center over time. They found that births, deaths, and international migration have all been instrumental in moving the center to the west over time. For example, there have been more births in the west than in the east, due largely to the relatively younger age structure in the west. Internal migration has generally acted to move the center westward, but there have been years (in the early 1990s, in particular) when the effect of migration alone (ignoring births, deaths, and international migration) has been – surprisingly – to move the center eastward. Thus the country's longstanding westward movement has slowed sufficiently that net movement is on balance no longer strongly oriented toward the west, and in fact internal migration has on balance occasionally been toward the east. In the early years of the second decade of the twenty-first century, net annual movement of the center that is attributable to internal migration has been on the order of one kilometer west, and one kilometer south. The center has also taken a sharp turn to the south in recent years. In the 19 years preceding 2014, in only 4 of those 19 years was the westward movement greater than the southward movement of the center.

3.1.3 Centers of Population for Large Geographic Regions

3.1.3.1 The Mean Center of Population as Calculated by the US Bureau of the Census

A starting point for thinking about the center of population for a large population residing on the surface of the earth is to take a weighted average of the latitude and longitude coordinates of locations, where the weights are again populations (which we will simplify, by assuming a sphere – accounting for the non-sphericity of the earth takes all of this to another level of sophistication, with little benefit in the sense that the location found with this added complexity would not be far away from the location found by ignoring it). This would not work, though, because although the length of a degree of latitude does not vary, the distance from one degree of longitude to the next varies with distance to the equator. Lines of latitude are always about 69 miles apart. Lines of longitude are maximally distant from one another at the equator (about 69 miles apart), and they converge at the poles.

To reflect this, the US Bureau of the Census, in its calculation of the center of population, uses a weighted average of latitudes. It also uses a weighted average of longitudes, where there is an additional weighting factor of the cosine of the latitude:

$$\bar{\phi} = \frac{\sum_{i=1}^{n} w_i \phi_i}{\sum_{i=1}^{n} w_i} ; \bar{\gamma} = \frac{\sum_{i=1}^{n} w_i \gamma_i \cos \phi_i}{\sum_{i=1}^{n} w_i \cos \phi_i} \qquad (3.5)$$

where ϕ_i and γ_i are, respectively, the latitude and longitude of location i. This can be understood (a bit!) by recognizing that the cosine of 0° is equal to 1 (and therefore, at 0° latitude at the equator, the population alone is the weight), and the cosine of 90° is equal to 0 (and therefore, at the poles – 90° latitude – where all points are in the same location regardless of longitude, the weight is equal to 0).

This method is equivalent to finding the center of population of locations plotted on a two-dimensional map that uses a Sanson–Flamsteed projection. More specifically, latitude–longitude pairs can be transformed to x–y pairs using $x_i = (\lambda_i - \lambda_0) \cos \phi_i$ and $y_i = \phi_i$, where the map is centered on the chosen longitude λ_0. If λ_0 is chosen to be the longitude of the center of population, the two-dimensional map will produce a center that is equal to the Census solution. Unfortunately, this projection does not preserve distances between points accurately, and hence the center of population as determined by the US Bureau of the Census is not a true center of gravity – it does not minimize the sum of squared great circle distances that people would have to travel to get there. In particular, as we shall see below, the location found by the US Bureau of the Census is about 1° of latitude (approximately 69 miles) to the south of where the actual center of population lies (see also Plane and Rogerson, 2015).

3.1.3.2 The Azimuthal Equidistant Projection

Barmore (1993: 8) began his engaging paper by relating how "I was flabbergasted when I discovered how the Bureau of the Census calculates the location of the center of population of the United States following each decennial census."

Barmore noted that if an azimuthal equidistant projection were centered on the center of population, distances and directions would not be distorted, and would be preserved accurately. After guessing the location of the mean center, point locations (e.g., country centroids) are first converted from latitude–longitude to x- and y-coordinates using the azimuthal equidistant projection centered on the location of the guess. Then the mean center is found for the x- and y-coordinates in the manner described above for the small-scale, two-dimensional case. Following this, the new mean center's x- and y-values are converted back to a latitude–longitude pair, and this constitutes the improved estimate of the location of the mean center. Finally, this improved estimate is used to repeat the steps just described. When there is minimal change in the location from one iteration to the next, the process is terminated. Equations associated with the projection as used in these steps are given in Appendix B.

Although Barmore does not explicitly point it out, an important characteristic of this mean center is that it is the point that minimizes the sum of squared great circle distances traveled from the n locations.

3.1.3.3 The Three-Dimensional Solution

Abofadel and Austin (2006) outlined an alternative method for finding the mean center for points on a sphere. First, for each location find

$$x = \cos \lambda \cos \varphi; \ y = \sin \lambda \cos \varphi; \ z = \sin \varphi \qquad (3.6)$$

where φ represents latitude and λ denotes longitude. Then the weighted mean center is found in the (x, y, z) coordinates in the usual way:

$$\bar{x} = \frac{\sum_{i=1}^{n} w_i x_i}{\sum_{i=1}^{n} w_i}; \ \bar{y} = \frac{\sum_{i=1}^{n} w_i y_i}{\sum_{i=1}^{n} w_i}; \ \bar{z} = \frac{\sum_{i=1}^{n} w_i z_i}{\sum_{i=1}^{n} w_i} \qquad (3.7)$$

The last step is to transform back to a latitude–longitude (and depth) for the mean center:

$$\varphi = \sin^{-1}(\bar{z}/t); \ \lambda = \tan^{-1}\left\{(\bar{y}/t)/(\bar{x}/t)\right\} \qquad (3.8)$$

where $t = \sqrt{\bar{x}^2 + \bar{y}^2 + \bar{z}^2}$. This gives the surface location of the solution; the actual center is below the surface, at a depth of $3{,}953(1-t)$ miles (3,953 miles is the radius of the earth).

To summarize, the latitude, longitude, and depth coordinates of the mean center are, respectively,

$$\sin^{-1}\left(\frac{\bar{z}}{t}\right), \ \tan^{-1}\left(\frac{\left(\frac{\bar{y}}{t}\right)}{\left(\frac{\bar{x}}{t}\right)}\right), \ (1-t)R \qquad (3.9)$$

where R is the radius of the earth. (Here we use negative longitudes for the US locations.) If the longitude result is less than -180, 180 is added to the result.

This method gives us a location that minimizes the (weighted) sum of squared distances from the locations on the surface of the earth, to the center (allowing for the fact that the center will be below the surface of the earth). As we will see, it comes close to minimizing the sum of squared great circle distances from locations to the mean center (which is on the surface of the earth), but does not quite achieve that objective.

3.1.3.4 The Median Center for Points on a Sphere

Finding the point of minimum aggregate distance (m.a.d.) has received relatively little attention for the case when points are located on a sphere. Litwhiler (1976) and Litwhiler and Aly (1979) suggested using the approach outlined in the previous section for the mean center – first project the locations into the two-dimensional plane using the azimuthal equidistant projection to preserve distance and direction, then find the median center for the two-dimensional data in the usual way, and then project back onto the sphere. Katz and Cooper (1980) formulated the problem of minimizing aggregate travel distance using great circle distances and they suggested using Weiszfeld's iterative approach. In what follows we make use of these ideas, outlining an explicit iterative method for finding the point of minimum aggregate great circle distances.

The sum of the great circle distances traveled from a set of points with latitude and longitude coordinates (φ_j, γ_j), $j = 1, 2, ..., n$, to the median center (φ_1, γ_1) is

$$S = R \sum_{j=1}^{n} \arccos[\sin \varphi_1 \sin \varphi_j + \cos \varphi_1 \cos \varphi_j \cos(\gamma_1 - \gamma_j)] \tag{3.10}$$

where (ϕ_j, γ_j) denotes the latitude–longitude pair for point j, and R is the radius of the earth. More generally, we can attach populations or weights (w_j) to each of the n locations and minimize

$$S = R \sum_{j=1}^{n} w_j \arccos[\sin \varphi_1 \sin \varphi_j + \cos \varphi_1 \cos \varphi_j \cos(\gamma_1 - \gamma_j)] \tag{3.11}$$

In the Appendix, Section A.1, we review the tools of calculus that may be used to find the minimum of this function. Here we state that the solution to this problem is governed by the following equations:

$$\varphi_1 = \arctan \frac{\displaystyle\sum_{j=1}^{n} \frac{w_j \sin \varphi_j}{\sqrt{1 - (\cos d_j / R)^2}}}{\displaystyle\sum_{j=1}^{n} \frac{w_j \cos \varphi_j \cos(\gamma_1 - \gamma_j)}{\sqrt{1 - (\cos d_j / R)^2}}} \tag{3.12}$$

and

$$\gamma_1 = \arctan \frac{\displaystyle\sum_{j=1}^{n} \frac{w_j \cos \varphi_j \sin \gamma_j}{\sqrt{1 - (\cos d_j / R)^2}}}{\displaystyle\sum_{j=1}^{n} \frac{w_j \cos \varphi_j \cos \gamma_j}{\sqrt{1 - (\cos d_j / R)^2}}} \tag{3.13}$$

We cannot solve these directly for the median center because we do not know the distances from the locations to the median center. We can, however, use these equations to carry out an iterative procedure to find the actual location of the median center. The first step is to make an initial guess at where it might be, say $\left(\varphi_1^{(0)}, \gamma_1^{(0)}\right)$. Then we compute the distances from this location to all n locations and use them on the right-hand side of the preceding two equations to produce a new median center location that serves as an update of the initial guess. Note that the updating equation for φ_1 requires not only these distances but also the current guess for γ_1, so that the latter quantity should be determined first. Next, we compute the distances from this updated estimate to all n locations and repeat the step of using the two equations to further update the coordinates of the median center. Convergence of the coordinates to a stable, unchanging location usually occurs after a small number of iterations.

Example 1

The latitudes and longitudes for Los Angeles, New York City, and New Orleans are (34.05°N, 118.25°W), (40.72°N, 74°W), and (29.95°N, 90.07°W), respectively. Assume equal numbers of people in each location and find the mean center (using census, three-dimensional, and azimuthal methods) and the median center (using the azimuthal projection). The solutions are given in Table 3.1. Examining the great circle distances of the three locations from each of the centers yields Table 3.2.

Table 3.1   Center of population for Los Angeles, New York City, and New Orleans using four different methods

	Latitude	Longitude	Depth (miles)
Azimuthal equidistant	36.281	94.586	
Three-dimensional	36.240	94.531	141.5
Census (Sanson–Flamsteed)	34.873	94.622	
Median center (azimuthal)	30.145	90.150	

Note: Assumes equal populations for the three cities

Table 3.2   Distances (miles) from the four different centers of population

	Azimuthal	Three-dimensional	Census	Median
Los Angeles	1340.15	1343.24	1342.17	1658.96
New York City	1149.91	1148.12	1190.99	1161.38
New Orleans	508.65	504.63	430.76	14.28
Total	2998.71	2995.98	2963.92	2834.62
Average distance	999.57	998.66	987.97	944.87
Standard distance	1061.01	1061.02	1065.46	1169.21

An interesting feature of the median center solution is that it is just 14 miles north of New Orleans. Note that the smallest aggregate travel distance is for the median center. Also reported is the standard distance; this is the square root of the average squared distance traveled. As expected among the alternatives, the mean center based on the azimuthal projection has the lowest sum of squared distances (and, equivalently, the lowest standard distance). The three-dimensional solution is an interesting one; it gives solutions that are very close to the "correct" solution provided by the azimuthal projection.

The previous example assumed equal populations. The 2017 populations for Los Angeles, New York City, and New Orleans were 4,000,000, 8,623,000, and 393,292, respectively. After weighting locations by population, the centers of population are as in Table 3.3.

Table 3.3   Weighted center of population for Los Angeles, New York City, and New Orleans using four different methods

	Latitude	Longitude	Depth (miles)
Azimuthal equidistant	40.411	-88.782	
Three-dimensional	40.512	-88.551	161.5
Census (Sanson–Flamsteed)	38.606	-88.889	
Median center	at New York City		

When one location has a population that is greater than the combined sum of all other populations, the median center is at that center with the greatest population.

The distances from each of the three places to the different centers are shown in Table 3.4. Table 3.4 also shows the average distance traveled and the standard distance associated with each center.

Table 3.4   Distances (miles) from the four different weighted centers of population

	Azimuthal	Three-dimensional	Census	Median
Los Angeles	1541.47	1530.50	1534.94	2442.68
New York City	901.28	912.44	918.90	0
New Orleans	724.62	732.81	600.13	1167.19
Mean distance	1104.13	1108.28	1114.10	774.05
Standard distance	1143.60	1144.98	1150.38	1375.03

As expected, these centers are located north of those that initially assumed equal populations, since the most southern city of New Orleans has a relatively smaller population than the other two places.

Note that use of the azimuthal projection minimizes the sum of squared distances from the three points to the center, while the median center point of minimum aggregate distance (m.a.d.) minimizes the sum of distances from the three points to the center.

───────────────────────────(**Example 2**)───────────────────────────

2010 Census data

Using county centroids and 2010 decennial county populations, national centers of population were derived for each of the methods (using the 48 continental states, plus the District of Columbia). They are shown in Table 3.5, along with the mean distance and mean squared distance traveled by individuals. The census center is 3 miles northeast of Plato, Missouri. The azimuthal center is about 8 miles north of Linn, Missouri; the Missouri River lies 10 miles to the northwest. The minimum aggregate distance (m.a.d.) center is east of the Mississippi River, about 5 miles southeast of Vincennes, Indiana.

Table 3.5 US centers of population (2010)

	Latitude	Longitude	Mean distance (m.)	Mean squared distance (mi)
Azimuthal	38.58	−91.70	806.3	829,383
Census	37.52	−92.17	815.0	835,259
m.a.d.	38.61	−87.40	784.9	882,939

Note that the mean center found using the azimuthal equidistant projection minimizes the sum of squared distances traveled, and the mean average distance traveled is lowest for the m.a.d. solution. For the median center, individuals are, on average, a little less than 800 miles away from the center – this gives us some idea of how spread out people are about the center. Note also that the US Bureau of the Census' center of population is considerably south of the other two. The census center is 77 miles from the center found using the azimuthal projection, and it is 270 miles from the m.a.d. solution. The azimuthal and m.a.d. solutions are separated by 232 miles.

For the azimuthal equidistant projection, the solution converges quite quickly on the final solution, even if the initial estimated location is not accurate. To illustrate, the projection was initially centered on the arbitrary and distant location of (47°N, 80°W); the updated center after the first iteration is then given as (38.71°N, 91.66°W), which is close to the actual solution.

3.2 Measures of Dispersion

Dispersion of the population is commonly measured by Bachi's standard distance (1963). It is equal to the square root of the average squared distance of individuals from the center of population: $\sqrt{\sum w_i d_i^2 / \sum w_i}$.

In the United States, Bachi's standard distance has increased from just under 300 miles in 1790 to close to 900 miles today, illustrating the large increases in the dispersion of the population around the center that have taken place.

3.2.1 The Standard Deviational Ellipse

The standard deviational ellipse, along with its associated parameters, constitute a more general method for describing dispersion.

Yuill (1971) gave a good description of the standard deviational ellipse as a summary measure for two-dimensional point distributions. Early explication of the concept is usually credited to Lefever (1926). First, the standard deviations along the x- and y-axes are found for a set of points scattered in two dimensions. Then the axes are rotated, and the standard deviations are found for the point coordinates along the new x- and y-axes. The angle of rotation that yields the largest standard deviation is used to create the new axes, and the length of the axis associated with that largest standard deviation serves as the major axis of the ellipse. Thus the orientation of the major axis of the ellipse is aligned with the direction of largest variability in point locations. The minor axis is perpendicular to the major axis, and its length is equal to the smallest of the standard deviations found through the rotation. The standard deviational ellipse thus provides a summary descriptor of the spatial distribution of points. The area of the ellipse should contain approximately 2/3 of the total number of points, and its eccentricity captures the degree to which the point pattern is roughly circular (low eccentricity) or strung out along a line (high eccentricity).

Gong (2002) noted that when the standard deviations are calculated as the axes are rotated, the locus of points that is traced out is actually *not* an ellipse. The actual shape (which Gong terms the "standard deviational curve" or SDC) is that of an hourglass, and so the corresponding ellipse fits exactly around the outer portion of the hourglass (which, then, is inscribed within the ellipse), with the major axis corresponding to a line through the major length of the hourglass, and the minor axis corresponding to a line drawn at right angles to the major axis, through the narrow portion of the hourglass.

For a set of n point locations (x, y), the angle of rotation yielding the maximum variance may be found as

$$\tan^{-1}\left(\frac{\sqrt{a^2 + b^2} - a}{b}\right) \qquad (3.14)$$

where $a = \left(\sigma_x^2 - \sigma_y^2\right)/2$ and $b = \left(\sum_{i=1}^{n} w_i x_i y_i\right)/\sum_{i=1}^{n} w_i$, and where σ_x^2 and σ_y^2 are the variances of the x- and y-coordinates, respectively. This assumes that the coordinates are expressed as deviations from the mean center (i.e., the origin is centered on the mean center), and thus $\sigma_x^2 = \sum_{i=1}^{n} w_i x_i^2 / \sum_{i=1}^{n} w_i$ (and the variance of the y's is found similarly). Associated with each of the n point locations may be a weight w_i representing population for example, as it is in the present application. The lengths of the axes are found as

$$\sqrt{\left(\sigma_x^2 + \sigma_y^2\right)/2 \pm \sqrt{a^2 + b^2}}$$ (3.15)

The orientation of the SDC in the United States has changed from 45° (northeast to southwest) in 1790, to close to 0° (i.e., a dominant east–west axis) where it has remained since the beginning of the twentieth century. Like Bachi's standard distance, the lengths of the axes have grown substantially over the course of two centuries, although that growth has slowed markedly in the last two decades.

3.3 Measures of "Coastiality"

Measures of "coastiality" – the degree to which people live near a coast – include the median distance the population resides from the coast (along with additional information on the distance distribution). To examine the population distribution along the coast in more detail, the shortest distance from each US county centroid to the coast was calculated. This was done by using ESRI's boundary file for North America, paring it down to include only those points ($n = 3,612$) along the east and west coast of the United States (and the Gulf of Mexico). The counties for the 48 continental US states and the District of Columbia ($n = 3,108$) were then ordered by distance to coast.

The median distance of the population (assigned to county centroids) to the coast was 129 miles in 2000; half of the population lived in counties with centroids that were closer than that to the coast. Between 2000 and 2009, there was a small decline in the proportion of people living within 20 miles of the coast, and a small increase in the proportion living within about 20–60 miles of the coast. By 2009 the median distance of the population from the coast had declined slightly, from 129 miles to 127 miles.

3.3.1 The Hu Line

The Heihe–Tengchong line (known as the Hu line; see, e.g., Naughton, 2007) is an imaginary line that cuts across China diagonally, from northeast to southwest. It divides China into two parts that differ markedly in terms of population density. It was suggested by Hu Huanyong in 1935, and at the time 4% of the population lived on 64% of the area west of the line; the remaining 36% of the area to the east contained 96% of the population. It is easily seen that associated with the original

Hu line is a type of Hoover index of concentration (in addition to the discussion in Section 3.5, see, e.g., Duncan et al., 1961; Long and Nucci, 1997), although the index has not been maximized – it is possible that by drawing the line in a somewhat different way, the Hoover index could be even higher. Population distribution with respect to this line has been very stable over time. The line in 2005 had 6% of the population living on 57% of the area to the west of the line, while 94% of the population lived on the 43% of the area to the east of the line (corresponding to a Hoover index of 57 – 6 = 51).

The Hu line is used here to find boundaries that maximize the difference between the high-density coastal population and the lower density inland population; it is also used here to find latitudes and longitudes that divide the country into high-density and low-density portions.

Both the proportion of total US population in 2010 and the proportion of total land area were cumulated for counties, in order of increasing distance from the coast. The maximum difference between these two quantities was 0.3041; this occurred when the 444 counties that had centroids that were closer than 53.4 miles to the coast were accumulated as a set. This group of counties contained 42.16% of the population and just 11.75% of the land area of the continental United States (plus DC). This suggests that one US version of the Hu line could be drawn in the form of two segments – one extending 53.4 miles inland from the east coast and Gulf of Mexico, and the other extending along the west coast from Washington to southern California, also at a distance of 53.4 miles from the coast. This index of concentration (when represented as a quantity between 0 and 100) has risen from just under 18 in 1910 to its current value of just over 30.

Alternative Hu lines – chosen to maximize the difference between cumulative population and cumulative area – are also readily found. A single Hu line, for example, can be chosen by trying different longitudes, and calculating the percentage population and percentage area to the east and west. For the continental United States, the Hu line is at longitude 90.515°W, traveling roughly through Wisconsin, Illinois, and loosely following the Mississippi River. The associated Hoover index is $H = 30.71$; to the west of the line lies 71.89% of the country's land area, but it contains just 41.18% of the population. The population density is 58.7 people per square mile to the west, and 214.2 people per square mile to the east.

The Hu latitude in 2010 was at 43.162°N; the Hoover index is 16.66, and this in turn results from the fact that 90.93% of the population resides on the 74.27% of the continental land area which is to the south of the line. The population density is 125.4 people per square mile south of the line, and just 36.1 people per square mile to the north of this latitude.

3.3.2 The Depth of the Three-Dimensional Center of Population

The depth associated with the center found by the three-dimensional spherical geometry approach can be interpreted as a measure of the dispersion of the population away from its center, or the bicoastiality of the population. Since the method is based upon the location of points on a sphere, the resulting mean center does not necessarily reside on the surface of that sphere.

Table 3.6 shows the third dimension of the mean center – the depth that the mean center lies below the earth's surface. We see from the table that this point has moved deeper below the surface over time and has just recently reached a depth of 100 miles. The movement of this mean center toward greater depths is understandable given that the country's population has spread out over a larger portion of the earth's surface over time. In fact, one can use this vertical distance below the surface as a measure of the nation's dispersion over space and specifically as a measure of its increasing tendency toward a bicoastal distribution.

Table 3.6 Measures of US bicoastiality

Year	% population in coastal states	Three-dimensional solution depth (miles)
1790	74.43	9.9
1800	70.53	11.4
1810	64.20	14.2
1820	58.02	16.7
1830	53.93	18.6
1840	46.74	21.9
1850	42.69	28.5
1860	38.55	37.4
1870	35.67	39.7
1880	34.00	45.1
1890	32.98	50.8
1900	33.05	52.5
1910	34.49	59.8
1920	35.22	62.7
1930	36.73	67.6
1940	37.63	70.3
1950	39.76	77.9
1960	41.64	84.1
1970	43.17	88.5
1980	43.26	92.5
1990	45.67	98.0
2000	45.98	99.4
2010	46.29	100.2

3.4 Geographic Centers

One of the first descriptive measures students of statistics learn about is the mean, or average. Adding the values of all observations together and then dividing by the number of observations to obtain the sample mean produces a measure of centrality that serves to describe the set of observations.

But how does one define the center of a geographic region? A geographic center of a region is the center of mass for that region – it is a balance point. If the region's shape is represented by a cardboard cutout, the geographic center is the point on which the cutout would balance. The geographic center (in two dimensions) is also an average of all of the *x*- and *y*-coordinates corresponding to all of the locations within the region.

For small geographic scales (e.g., within a city or county), the solution to the problem of finding the geographic center of a two-dimensional polygon, as we will see, may easily be found by using *x*- and *y*-coordinates of the vertices. The solution is more complex when the problem involves large geographic scales that necessitate consideration of the curvature of the earth (see, e.g., Rogerson, 2015a).Where is the center of the North American continent? And, better yet, why does it matter (if at all!)? There have been many claims to different locations over the years, and there have been just as many disputes. The location is important in part because of tourism – people do travel to these places – and also because it is, to some extent, a matter of civic pride and identity. Geographic centers of states and countries are often marked by monuments, and some people (many, or perhaps very few?) organize trips and vacations to include visits to these locations, just as others may have the goal of traveling to the capital, or high point, of each state.

Geographic centers of counties also had some historical importance in the United States – they were often mandated as the locations for county court houses and governmental offices, since this would serve to maximize accessibility from all points in the county.

Finding the center of geographic areas would seem to be a basic calculation that geographers could master. Yet so many questions arise, and so little careful attention has been given to the topic, that debates about actual locations rage on.

In 1930, Bulletin 817 written by Edward Douglas and published by the US Geological Survey (USGS) stated that the center of North America was at 48.17°N, 100.17°W – 16 miles southwest of Rugby and not far (5 to 6 miles) from the towns of Orrin and Balta. Because Rugby was the county seat, a monument was located there in 1932. It still stands and attracts many visitors each year. Rugby trademarked the title "Geographical Center of North America" and Rugby's location has been listed in virtually every almanac and statistical abstract since. An alternative center was constructed near Orrin by two brothers (Joseph Buechler and Wendelin Bickler; they couldn't agree on how to spell their last name) that served the dual purposes of a monument for the center and a shrine to the Virgin Saint Mary.

To find the center, Douglas used a cardboard cutout of North America and dangled it with a string attached to various parts of the periphery, each time hanging a plumb bob and drawing a line on the cardboard. The intersection of these lines gave him the location. A similar method for finding geographic centers used by USGS at the time was to mark the center as that point where the cardboard cutout balanced on the tip of a pin point. A letter to the *Argus Leader* (Sioux Falls, SD; Oct. 2 1932) referred to two separate letters by the USGS in response to queries about their method – the Director himself referred to the plumb bob method, and another response from USGS noted that the map had been balanced on a pin point. The agency has gone to great pains over the years to note that the method is not exact, and in fact it has gone even further at times to say that there is no single method that can be used. Indeed, in giving the center for North America and the individual states, the original publication notes that "[t]he exact position of the center ... cannot be determined from the data available, but the following approximate positions are sufficiently exact for ordinary purposes." Despite this, the centers determined in this manner have withstood the test of time. The need for identification with such places, as well as the fact that they have not been determined more precisely, are both remarkable.

Statistical abstracts and almanacs have perpetuated the list of geographic centers for states since they were first found by these crude methods in the 1930s. We should be able to do better, and in this section we will look at a method for finding the geographic center that uses a boundary file for a region along with (a) methods that find the geographic center of a two-dimensional polygon and (b) generalizations of this that account for the curvature of the earth.

3.4.1 Geographic Center for Two-Dimensional Polygons

An algebraic method for finding the geographic center is well known for two-dimensional polygons represented on a flat surface.

For a two-dimensional representation of a geographic area, the center may be found by averaging all of the x- and y-coordinates within the region, but to do so in practice requires integrating over all of the infinite number of locations within the region. For a two-dimensional polygon, a direct solution for the centroid (C_x, C_y) of a polygon with vertices (x_0, y_0), (x_1, y_1), ..., (x_{n-1}, y_{n-1}) may be found as follows (Bashein and Detmer, 1994):

$$x_c = \frac{1}{6A}\left\{\sum_{i=0}^{n-1}(x_i + x_{i+1})(x_i y_{i+1} - x_{i+1} y_i)\right\}$$

$$y_c = \frac{1}{6A}\left\{\sum_{i=0}^{n-1}(y_i + y_{i+1})(x_i y_{i+1} - x_{i+1} y_i)\right\}$$

$$(3.16)$$

The area, A, of the polygon is found as

$$A = 0.5 \left\{ \sum_{i=0}^{n-1} x_i y_{i+1} - x_{i+1} y_i \right\} \tag{3.17}$$

and it should be noted that the n coordinate pairs are numbered sequentially and counterclockwise as they occur along the perimeter, beginning with 0 and ending with $n - 1$. Additionally,

$$(x_n, y_n) = (x_0, y_0) \tag{3.18}$$

If the pairs are numbered in clockwise order, the area will be negative, but the coordinates of the centroid will still be correct.

3.4.2 Geographic Centers for the Three-Dimensional Case

The solution is more complex when the region is on the surface of a three-dimensional sphere. Here one would like to find the point that minimizes the sum of squared great circle distances from it, to all points in the region.

Boundary files consist of a list of latitude and longitude coordinates for consecutive vertices of a polygon that represents the boundary. A two-dimensional map can be made, and the known techniques for finding the geographic center can be used. But of course there are many choices of map projection in making two-dimensional maps.

The key to finding the true geographic center for a large region is to use an azimuthal equidistant projection – this projection preserves distance measurements, as long as the distance is measured from the center of the map. But of course we don't know the center! The procedure then is to first guess where the geographic center is. A two-dimensional polygon map is made using the azimuthal equidistant projection (using the equations in Appendix B). Then the geographic center of this two-dimensional polygon is found. Coordinates are next converted back to latitude and longitude (again using the equations in Appendix B). This point is used to create a new azimuthal equidistant map with x- and y-coordinates, serving as the center of the new map. After several iterations, the points will converge on a single solution.

The algorithm for finding the geographic center (or center of mass) for a region on the earth's surface may therefore be summarized as follows:

0. Guess where the center is (in latitude and longitude coordinates).
1. Convert latitude and longitude coordinates of the boundary points into x- and y-coordinates using the formulas for the azimuthal equidistant projection (see equations in Appendix B).
2. Find the geographic center for the two-dimensional polygon using Equations 3.16–3.18.
3. Convert the center found in step 2 back into latitude and longitude coordinates using the equations in Appendix B. This is the new and improved estimate of the geographic center.

4. Use the improved estimate from step 3 and return to step 1.
5. Iterate steps 1–4, and stop when the estimate does not change (or does not change much).

Of course the definition of the boundary can make a difference. Using the NOAA-defined boundary (www.ngdc.noaa.gov/mgg/shorelines/ [accessed September 11, 2020]) based on the 33,208 boundary points that exclude the Great Lakes (because the St. Lawrence River is the outlet for the Great Lakes, and its mouth in this dataset is not part of the boundary) leads to the geographic center of North America at 47.1760°N, 101.3021°W, a few miles north of Center, North Dakota. Including the Great Lakes by drawing the boundary across the mouth of the St. Lawrence (along with a few less significant differences), and using the ESRI boundary file data, the location of the center changes to 47.6132°N, 100.7750°W based upon 15,654 boundary points. This location is about 12 miles north of Mercer, North Dakota, and about 39 great circle miles (but 63 driving miles) northeast of Center. It is also roughly 89 driving miles south of Rugby.

McCollum (2011) carefully considered the location of the geographic center of North America, using the iterative procedure described above, together with a mixture of the solutions obtained using azimuthal equidistant and Lambert azimuthal projections (the latter preserves area, but not direction). He also considered six different definitions for the North American continent. For essentially the same definition described in the previous paragraph (including the Great Lakes), McCollum found that the center was at 47.609°N, 100.74°W, also near Mercer, and less than 2 miles away from the slightly more accurate calculation that relies on the azimuthal equidistant projection alone and noted above. McCollum's location is about 63 miles southwest of Rugby, and a little more than halfway on the route from Rugby to Center.

3.4.3 Geographic Center of the Continental United States

The geographic center of the continental United States has historically been located in Lebanon, Kansas. Using the azimuthal equidistant projection places the center in Agra, about 26 miles to the west. (An interesting aside is that the town of Agra was named after the city of Agra in India, where the Taj Mahal is found.) In 1959, after the addition of Alaska and Hawaii as states, the National Geodetic Survey found a point about 20 miles to the north of Belle Fourche, SD as the geographic center of the United States. A monument was erected in 2007; it is a bit over 400 miles SSW from Rugby.

3.4.4 Geographic Centers of Continents

Table 3.7 gives the latitude and longitude coordinates for the geographic centers of all seven continents, based upon the azimuthal equidistant projection and boundary files

that are publicly available and downloadable from ESRI (arcgis.com/home/item.html?id
=a3cb207855b348a297ab85261743351d [accessed September 11, 2020]). Some comments about the centers for some of the continents follow.

Table 3.7 Geographic centers for continents

N. America	47.6132°N, 100.7750°W	12 miles north of Mercer, North Dakota
S. America	13.8844°S 60.4660°W	Bolivia; < 100 feet from river boundary with Brazil
Asia	47.8876°N, 95.2432°E	35 km northeast of Durvuljin, Mongolia; 267 miles south of Kyzyl
Europe	54.4853°N, 26.5138°E	Belarus, about 113 miles southwest of current center in Polotsk
Africa	6.908°N, 18.0762°E	
Antarctica	83.4877°S, 101.2619°E	
Australia	27.6466°S, 134.0297°E	About 20 miles southeast of Marla, South Australia

3.4.4.1 Center of Asia

Tuva is a Russian republic; it is perhaps best known as the destination that was a longstanding quest for the Nobel physicist, Richard Feynman. It lies at or near the geographic
center of Asia, and discussion of the center is prominent in Ralph Leighton's (1991)
account of his joint efforts with Feynman to get to Tuva. In a book written in Russian
(Promptov, 1950), it is noted that two lines connecting the outermost points of a map of
Asia (one running from north to south, and the other running from east to west) inter-
sect at Tuva. There is reference to an unknown traveler who had already visited the center
of two other continents, and who had succeeded in leaving a crude monument in Tuva.
Since 1968, there has been an obelisk in the capital city of Kyzyl marking and/or symbol-
izing the location (at 51.66°N, 95.92°E). Other sources corroborated the account (e.g.,
Leonov, 1927). Not putting complete faith in Promptov's method for finding the center,
Leighton next decided to find the center of Asia by balancing a map on the head of a pin.
He tried several different projections until he found one that worked – Gall's stereo-
graphic projection. To "confirm" the accuracy of the method, he used a cutout map of
North America, using the same projection. Hoping that Rugby would be the location of
the balance point, he instead found a location to the north, in Saskatchewan. A librarian
at the location in Saskatchewan gave him some vague confirmation that there used to be
some claim of that location as the center. Closer to the subject at hand, Leighton related
how Feynman gave him a formula involving cosines and sines; after using it together
with several hundred boundary points input by hand, Leighton found the center of Asia

at a point (45.52°N, 86.98°E) more than 500 miles southwest of Kyzyl, the capital of the Tuvan republic.

Some 700 km to the south of Kyzyl, in Yongfeng, China there is a second monument marking the geographic center of Asia. This is at 43° 40′ 52″N, 87° 19′ 52″E (about 25 km from Urumqi) and has been referred to as the Geographic Center of Asia – The Heart of Asia.

Application of the iterative azimuthal equidistant projection method and the boundary file for Asia supplied by ESRI gives the center of Asia at 47.8876°N, 95.2432°E. This is about 35 km (20 miles) northeast of Durvuljin, Mongolia and 267 miles close to due south of Kyzyl. It is also a bit more than 400 miles north and east of the other center, near Urumqi.

3.4.4.2 Comments on Centers of Other Selected Continents

The geographic center of Australia is just off route A87 and 142 miles almost due south of the widely acknowledged center at 25°36′36.4″S and 134°21′17.3″E, about 200 km south of Alice Springs – Lambert Centre. See, for example, www.ga.gov.au/scientific-topics/national-location-information/dimensions/centre-of-australia-states-territories #fivecentres (last accessed October, 2020)

South America: The center noted in the table is based upon the ESRI boundary file. The location is in Bolivia, less than 100 feet from the banks of the meandering Rio Itenez O Guapore River that serves as the border with Brazil. An alternative boundary file (from NOAA's shoreline dataset at www.ngdc.noaa.gov/mgg/shorelines/_ [last accessed October, 2020]) yields a center at 13.8296°S, 60.4926°W, a difference of just 3.8 great circle miles from the center noted above. This center is still in Bolivia and is less than a mile from the border with Brazil. There are of course alternatives for the geographic center that have been claimed over the years. There is a white marble obelisk in Cuiaba, Brazil (320 great circle miles to the east-southeast of the location above), with an inscription marking the location as the "geodesic center of South America." A new claim in the 1990s placed the center near Chapado dos Guimarães, about 28 miles to the northeast of Cuiaba.

Africa: The geographic center is 327 miles north of the point described in the table below, and a little over 10 miles over land, northwest of Bagou, in the northern part of the Central African Republic. See, for example, worldatlas.com/webimage/countrys/aflan-dst.htm which has the center in Lobeke Natl Park has 2.62N, 16.10E° and www.iol.co.za/motoring/latest-launches/holgates-quest-the-heart-of-africa-1893745

Table 3.8 gives the geographic centers of selected countries, again found using the azimuthal equidistant projection and the ESRI boundary files. (These countries were selected in part because of previous attempts to designate a geographic center in the country. The interested reader may wish to investigate further how well the locations in Table 3.8 match those previous efforts.)

Table 3.8 Geographic centers of selected countries

	Latitude	Longitude
Hungary	47.1704°N	19.3974°E
Ireland	53.1667°N	8.1544°W
Lithuania	55.3323°N	23.919°E
Norway	63.7845°N	12.2293°E
Poland	52.1141°N	19.4238°E
Serbia	44.2156°N	20.7992°E
Slovenia	46.1248°N	14.8142°E
Switzerland	46.8021°N	8.2278°E
Great Britain	53.8320°N	2.4268°W
England	52.5629°N	1.4635°W
Northern Ireland	54.6110°N	6.6931°W
Scotland	56.6653°N	4.0165°W
Wales	52.2955°N	3.7419°W

3.5 Measures of Inequality

The Lorenz curve (1905) provides a visual depiction of inequality; it consists of a plot of cumulative values on both the vertical and horizontal axes. In its most common application, namely to income inequality, incomes are first ranked from lowest to highest (most often in groups, or categories, since individual data are not usually available) and then cumulative population is plotted on the x-axis, and cumulative income on the y-axis. The degree to which the curve bows away from the 45° line is a measure of inequality.

The Lorenz curve can also be used in the context of population distribution, where regional populations are first ordered in terms of increasing population density, and then cumulative land area is plotted on the horizontal axis, against cumulative population on the vertical axis. The Lorenz curve in Figure 3.1, based on the populations and areas of US counties in 2010, shows for example that 54.2% of the land area contained just 2.6% of the population (see Rogerson, 2019).

There are at least two measures related to the Lorenz curve that provide a numerical assessment of inequality. The Hoover Index of Concentration (Hoover, 1941) is the largest vertical distance between the Lorenz curve and the 45° line. It is also known as the Pietra index or ratio (the index has its antecedents in the work of Pietra), the Robin Hood index, and the Schutz index or Schutz coefficient. It can be interpreted – in the case of population – as the proportion of people who would have to move from high-density

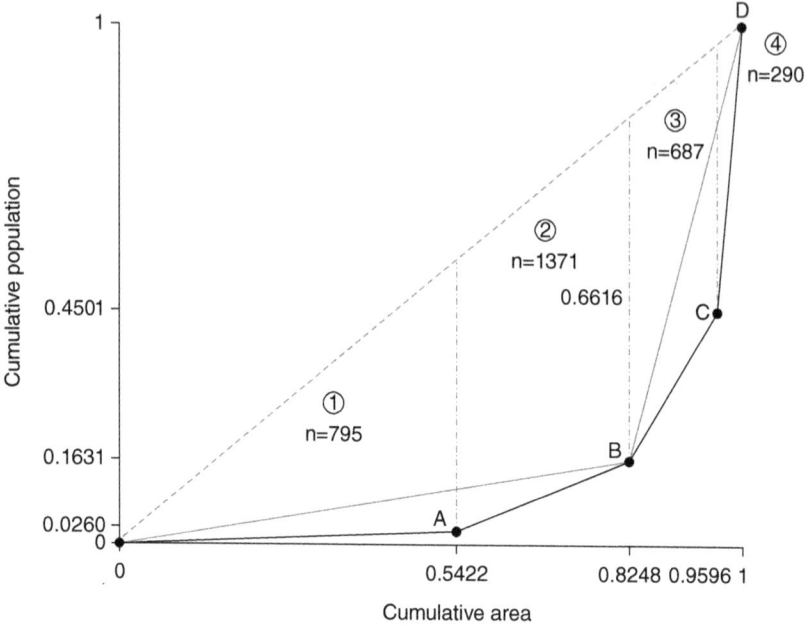

Figure 3.1   Lorenz curve for US population: 2010
Source: Rogerson (2019). Reprinted with permission of Springer Nature.

areas to low-density areas, to equalize population density in all regions. It may be calculated as half of the *relative mean absolute deviation* (from the mean):

$$H = \frac{\sum |x_i - \bar{x}|}{2n\bar{x}} \tag{3.19}$$

For US counties in 2000, the Hoover index was 65.61. Slightly more than 65% of the population would have to move from the high-density counties to the low-density counties
to equalize population density across the country. The distribution of population at the
county level became increasingly concentrated over the next 15 years, with the index
increasing to 66.16 in 2010, and 66.73 in 2015 (Rogerson, 2019).

    A second numerical measure of inequality is the Gini coefficient (Gini, 1912, 1914; for
more on the origins, see Ceriani and Verme, 2012). It is equal to the proportion of area
below the 45° line that lies between the 45° line and the Lorenz curve. In the case of
income inequality, it is also equal to half of the mean absolute difference between all
pairs of income:

$$GC = \frac{\sum_{i=1}^{n}\sum_{j=1}^{n} |x_i - x_j|}{2n^2\bar{x}} \tag{3.20}$$

Thus it can be interpreted as the expected proportion of mean income that would have to be given from one individual to another, if two random individuals met and decided to equalize their incomes. The Gini coefficient is always at least as high as the Hoover index.

The Lorenz curve is constructed from cumulative values; in the case of population, we use cumulative area on the x-axis and cumulative population on the y-axis. The four rows in Table 3.9 give these values for the US population in 2010. The cumulative areas and cumulative populations in each row constitute, respectively, the x- and y-values for a coordinate pair that can be plotted on the graph. The Lorenz curve made from this table will therefore consist of 4 points along the curve (along with the coordinate pairs $(0, 0)$ and $(1, 1)$), and these 6 points can each be connected with straight lines.

Table 3.9 Areas and populations for US counties, 2010

Density Group	Proportion of total population (p)	Proportion of total area (a)	Cumulative population (cp)	Cumulative area (ca)	Absolute difference ($cp - ca$)
1	0.0260	0.5422	0.0260	0.5422	0.5161
2	0.1371	0.2824	0.1631	0.8248	0.6617
3	0.2970	0.1348	0.4501	0.9596	0.5095
4	0.5499	0.0404	1.0000	1.0000	0.0000

Source: Rogerson (2019). Reprinted with permission of Springer Nature.

Rogerson (2013) showed that the Lorenz curve can be disaggregated into a set of further Lorenz curves (dubbed here sub-Lorenz curves) that depict inequality along the primary curve (see Figure 3.1). Thus the set of counties with low population density and the set of people with low incomes have Lorenz curves of their own, with accompanying Hoover indexes. Rogerson showed how the Gini coefficient can be written as a weighted sum of the Hoover indexes along the Lorenz curve. For example, if the data for US counties in 2010 are taken as shown in Table 3.9, we see that the Hoover index is 0.6617, since 16.31% of the population lives on 82.48% of the land area. (This was arrived at by looking at a list of counties ordered in terms in increasing density, and finding the breakpoint where the difference between cumulative population and cumulative area was a maximum.) When there are just two groups of counties (high density and low density), the Gini coefficient is equal to the Hoover index, and it is equal to the area of triangle OBD (as a proportion of the area under the 45° line) in Figure 3.1. Triangles OAB and BCD represent the inequality of population density *within* these low and high population density groups, respectively. The vertical heights of these triangles are essentially scaled versions of the Hoover indexes that could be calculated for the sets of low and high population density regions. These vertical lines divide the low-density regions into two subgroups (low–low, and high–low) and also divide the high-density regions into two subgroups (low–high and high–high). The Gini coefficient for these four groups of regions (corresponding also to the groups in Table 3.9) may be calculated as a weighted sum of these vertical distances which are scaled Hoover indexes, where the weights are the proportions of total land area that is associated with each section of the Lorenz curve. The vertical distances are calculated using the data in Table 3.9. For triangle OAB, the scaled Hoover index is

$$(0.1631/0.8248)(0.5422) - 0.0260 = 0.1072 - 0.0260 = 0.0812 \qquad (3.21)$$

For triangle BCD, the vertical distance and scaled Hoover index is $0.1631 + (0.9596 - 0.8248)(1 - 0.1631)/(1 - 0.8248) - 0.4501 = 0.8070 - 0.4501 = 0.3569$. In general these two vertical distances are calculated for triangles OAB and BCD respectively as

Area of OAB $= cp_2/ca_2(ca_1) - cp_1$ and
Area of BCD $= cp_2 + (ca_3 - ca_2)(1 - cp_2)/(1 - ca_2) - cp_3$ $\qquad (3.22)$

where cp_i and ca_i are the cumulative population and areas for group i, respectively (where the four groups are labeled in terms of increasing population density, as in Table 3.9).

The Gini coefficient is the weighted average of these vertical distances:

$$GC = 0.6617 + (0.8248)(0.0812) + (0.1752)(0.3569) = 0.7912 \qquad (3.23)$$

where the weights of 0.8248 and 0.1752 are the proportions of land areas associated with the left and right portions of the Lorenz curve respectively.

The second and third terms in the sum in (3.23) also represent twice the areas of triangles OAB and BCD in Figure 3.1, respectively. It is easy to see that the Gini coefficient is always at least as high as the Hoover index.

The Hoover indexes associated with the two sub-Lorenz curves can be found from

$$H_L = \frac{ca_1}{ca_2} - \frac{cp_1}{cp_2} \qquad (3.24)$$

$$H_U = \frac{ca_3 - ca_2}{1 - ca_2} - \frac{cp_3 - cp_2}{1 - cp_2} \qquad (3.25)$$

where H_L and H_U denote the Hoover indexes associated with the two groups comprising the lower (left) and upper (right) portions of the Lorenz curve, respectively.

For the 2010 census data in Table 3.9, $H_L = 0.5422/0.8248 - 0.0261/0.1631 = 0.4974$ and $H_U = (0.9596 - 0.8248)/(1 - 0.8248) - (0.4501 - 0.1631)/(1 - 0.1631) = 0.4265$.

These values allow us to look at inequality along the curve. Systemwide inequality may be disaggregated into inequality within subgroups (e.g., inequality of population density within low-density counties, or the inequality within particular income groups, such as the low-income population).

For the US population, during the first part of this century the low-density counties have become more concentrated; the group of counties with the highest density have deconcentrated somewhat; people are now spread out geographically a bit more across these counties.

Illustration: Income data

US individual income distribution data are available for single percentiles at https://dqydj.com/income-percentile-calculator/ (accessed September 12, 2020). These data in turn are taken from the IPUMS Current Population Survey data at https://cps.ipums.org/cps/ (accessed September 12, 2020; see Flood et al., 2020).

We now explore this data for 2016, 2018, and 2019.

Analysis of income inequality for 2018 begins here by finding the cumulative distribution of income. For example, the incomes corresponding to the lowest 1%, 2%, 3%, …,10% of individuals was, in 2018: $0, $0, $51, $922, $2,015, $2,808, $4,097, $5,121, $6,145, and $7,169. The last percentile reported (99th) was $307,831. For each percentile, we can calculate the total income accruing to everyone making that much money, or less. Supposing that there are 99 people in a hypothetical population: adding the ten numbers above yields $28,328. Adding all 99 numbers together yields $5,310,062. Thus the lowest 10% of the population has 28,328/5,310,062 = 0.00533 (about half of 1%) of the total income. Carrying this out for all percentiles reveals that the maximum difference between cumulative percent population and cumulative percent income is the Hoover index of 0.3270; 63% of the population had 30.30% of the income in 2018.

Taking the low-income group as this 63% of the population, we can examine income inequality *within* that group. The Hoover index for that group is 0.2489; 47.62% of the population in that group earned just 22.73% of the group's income. On the overall Lorenz curve we find that this corresponds to the 0.4762(63) = 30% of individuals who made 6.89% of the total income; the vertical distance on the lower sub-Lorenz curve at an x-value of 0.30 is (0.303/0.63(0.30)) - 0.0689 = 0.0754.

Similarly, we can examine income inequality within the remaining 100 - 63 = 37% of the population that earned high incomes. There we find from the data that, within the group of high earners, 66.67% of the group earned just 47.54% of the group's income, yielding a Hoover index of 0.1913. On the overall Lorenz curve we find that 63 + 0.667(36) = 87% of individuals made 63.43% of the income; the vertical distance on the upper sub-Lorenz curve at an x-value of 0.87 is 0.303 + (0.87 - 0.63)(1 - 0.303)/(1 - 0.63) - 0.6343 = 0.1208.

(Continued)

The Gini coefficient corresponding to these four income groups (low–low, low–high, high–low, and high–high) is the weighted average

$$GC = 0.3270 + 0.63(0.0754) + 0.37(0.1208) = 0.3270 + 0.0475 + 0.0447 = 0.4192 \qquad (3.26)$$

For 2019,

$$GC = 0.3278 + 0.63(0.0636) + 0.37(0.1236) = 0.3278 + 0.0401 + 0.0457 = 0.4136 \qquad (3.27)$$

and for 2016,

$$GC = 0.3270 + 0.63(0.0767) + 0.37(0.1224) = 0.3270 + 0.0483 + 0.0453 = 0.4206 \qquad (3.28)$$

A similar analysis for the other two years results in the summary table below. For the Hoover indexes associated with the lower and upper portions of the Lorenz curve, two numbers are given. The first is the Hoover index for that subgroup of the population; the second is the vertical distance from the original Lorenz curve to the diagonal lines drawn to create the sub-Lorenz curves.

Table 3.10 shows that income inequality, as measured by the Hoover index, remained the same from 2016 to 2018, and then increased slightly. The Gini coefficient tells a slightly different story, decreasing throughout the period. The table reveals that the decline is due primarily to decreases in inequality among those in the lower portion of the income distribution. Inequality among those in the upper portion declined a small amount by 2018, but then increased in 2019.

Table 3.10 Hoover and Gini indexes, 2016–19

	Median income	Hoover	Hoover (lower)	Hoover (upper)	Gini
2016	37610	0.3270	0.2532; 0.0767	0.1937; 0.1224	0.4312
2018	39990	0.3270	0.2489; 0.0754	0.1913; 0.1208	0.4295
2019	40100	0.3278	0.2302; 0.0636	0.1952; 0.1235	0.4136

Data source: https://dqydj.com/average-median-top-individual-income-percentiles/, based on data from the Current Population Survey, US Bureau of the Census

3.5.1 Measures of Inequality in Physical Geography and Other Fields

The study of topography has included the study of the "roughness" of the landscape, and this has entailed measuring the variability in elevations.

Grohmann et al. (2011) note that there are several definitions of surface roughness; they use the idea that it is the "variability of a topographic surface, *at a given scale.*" Olaya (2009) related that it "indicates how *undulating* the terrain is, i.e., how complex it is." Commonly used measures include the standard deviation of elevation values, and the standard deviation of the residuals derived from fitting a plane through the elevation values. Melton (1965) as cited in Olaya (2009) measured "ruggedness" as the range divided by the square root of the area.

This has been of particular interest in the study of drainage basins. A well-known measure used in this context is the *hypsometric integral.* Conceptually, the hypsometric integral is the proportion of a rectangular profile of land that is "remaining" after erosion has partially completed its work of taking all of the land in the area down from its maximum elevation to its minimum elevation. The **hypsometric curve**, in turn, is a plot of elevation on the *y*-axis vs. cumulative area at or above that elevation, on the *x*-axis. For example, about 20% to 25% of the earth's surface area is at sea level, or above. Elevation is sometimes expressed as a fraction of maximum elevation, and when it is, the hypsometric integral is the area under the curve (see Figure 3.2 for an example of a hypsometric curve). Pike and Wilson (1971) showed that the hypsometric integral can be calculated simply as the ratio

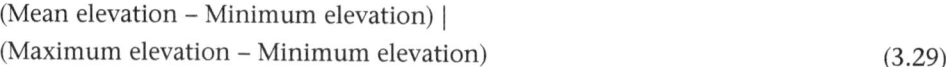

(Mean elevation – Minimum elevation) |
(Maximum elevation – Minimum elevation)   (3.29)

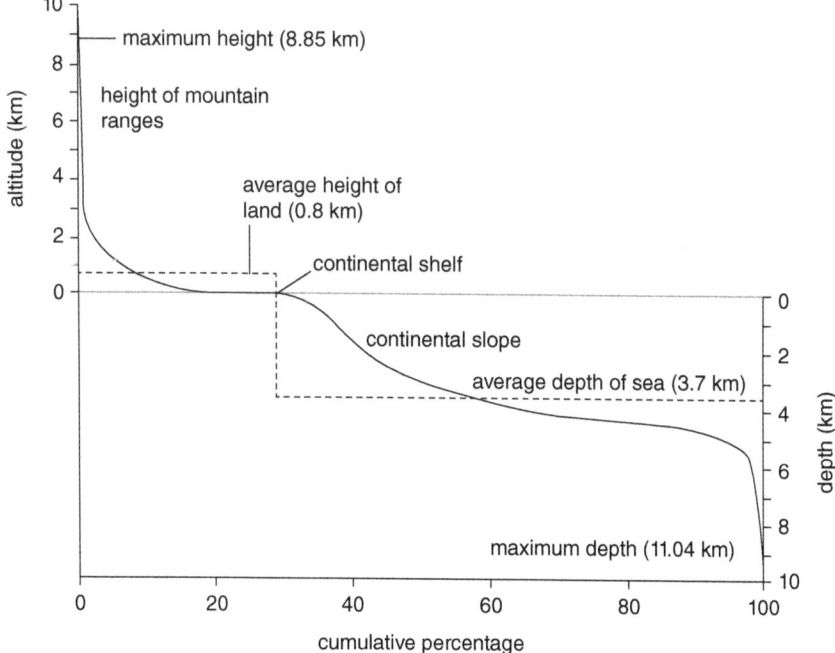

Figure 3.2   Hypsometric curve

Source: https://en.wikipedia.org/wiki/Hypsometry

This is equivalent to

{Mean / Range – Minimum / Range}   (3.30)

The first term of (3.30) is the substantive core of the measure. Since the second term serves as a reference point, note that the hypsometric integral is essentially a ratio of the mean to the range – it is a ratio of a measure of central tendency to a measure of dispersion. The first term is similar to the inverse of the coefficient of variation (standard deviation/mean). Many measures of inequality are variants of the coefficient of variation. This suggests that, since the hypsometric integral relies on an overly simplistic measure of dispersion that is based upon just two data points (i.e., the range), an alternative version of the hypsometric integral might be based upon the ratio of the mean to the standard deviation (which is the inverse of the well-known coefficient of variation ($s/\bar{x}$)).

Since many measures of inequality are unitless relative measures of variability, it is appropriate here to consider robust measures that rely on more than just the range of data (since the range is entirely dependent upon just two values – the minimum and the maximum), but less than a full distributional assumption. The interquartile range, which is the difference between the third and first quartiles (or 75th and 25th percentiles) in a dataset, is an example of a robust measure of absolute deviation. It is equivalent, visually, to the length of the rectangle in a boxplot. Occasionally, the semi-interquartile range (Q3 – Q1)/2 is used instead.

A robust measure of *relative* deviation is the unitless *quartile coefficient of dispersion*. It is equal to the semi-interquartile range, divided by the average of the first and third quartiles:

(Q3 – Q1) /2) / (Q1 + Q3) / 2) = (Q3 – Q1) / (Q3 + Q1)   (3.31)

This measure appears to be used often in biostatistics and health studies (see, e.g., Mariappan, 2013; Cysarz et al., 2015; Salinas et al., 2015). This measure suggests another potential robust measure of relative dispersion – the interquartile range, divided by the median, that is, (Q3 – Q1)/Q2. This is similar to the quartile coefficient of dispersion, and it is a robust alternative to the coefficient of variation. The inverse of this could for example be used as the basis of the hyspometric integral. A more detailed discussion of such measures can be found in Arachcige et al. (2019).

─────────────────────────────(**Example 1**)─────────────────────────────

Harlin (1978), in the course of his focus on properties of the hypsometric curve, provided an illustration for several eroded upland areas of Iowa and Nebraska. Figure 3.3 shows the topographical sheet from Glenwood, Iowa as an example. Table 3.11 compares the hypsometric integral for four areas with the Hoover index and Gini coefficient derived for those same areas. As expected, there is a negative relationship between the hypsometric integral and these other measures – as the amount of land remaining decreases, the hypsometric integral is lower (by definition), and the amount of inequality in elevations (as measured by *H* and *GC*) increases. The negative correlation of the hypsometric integral is slightly higher in absolute value with the Gini coefficient than it is with the Hoover index. A more detailed table is shown in Table 3.12.

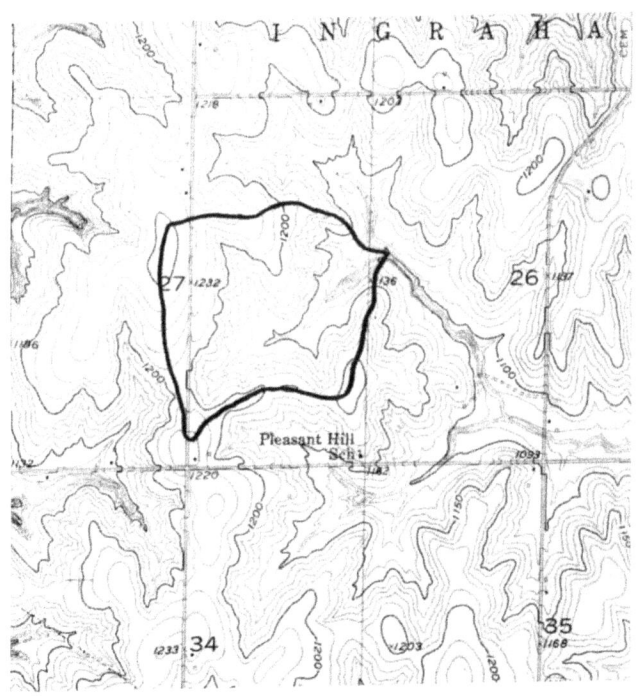

Figure 3.3   Topography of Glenwood, Iowa

Source: Harlin (1969). Reprinted with permission of Springer Nature.

Table 3.11   Hypsometric integral and Hoover and Gini indexes for four geographic areas

	Mean elev.	Min elev.	Max elev.	Hypsometric integral	Hoover index	Gini coefficient
Wolbach, NE	1185	1120	1248	50.81 (47.2)	19.11	26.23
Glenwood, IA	1166	1097	1216	57.77 (55.9)	17.32	23.42
Nickerson, NE	1319	1259	1372	53.39 (54.5)	17.86	25.04
Legrand, IA	960	913	984	66.29 (71.0)	14.63	20.27

Notes: Hypsometric integrals calculated from overlaying a 200-point grid on study area. Integrals in parentheses are from Harlin (1978)

Table 3.12 Topographic measures for four geographic areas

	Glenwood	Nickerson	Legrand	Wolbach
min	1097	1259	913	1865
max	1216	1372	984	2031
s_y	28.41	26.66	17.29	35.71
xbar y	1166	1319	960	1932
xbar r	69	60	47	67
coef var y	0.0244	0.0202	0.0180	0.0185
coef var r	0.4121	0.4407	0.3673	0.5344
Hypso. Integral	0.5777	0.5340	0.6629	0.4041
iqr/(Q3 + Q1) y	0.0199	0.0144	0.0120	0.0125
iqr/(Q3 + Q1) r	0.3451	0.3111	0.2353	0.3877
Hoover	0.1732	0.1786	0.1463	0.2124
Gini	0.2342	0.2504	0.2027	0.3008
% area	0.445	0.450	0.415	0.565
% elev	0.272	0.2714	0.2687	0.3527
# pts (of 200)	89	90	83	113
Hoover (low elev)	0.1725	0.2028	0.1889	0.1877
% area	0.4944	0.4222	0.4458	0.4602
% elev	0.3219	0.2194	0.2569	0.2725
# of pts	44	38	37	52
Hoover (high elev)	0.0582	0.0719	0.0503	0.1069
%area	0.5405	0.5455	0.4701	0.5517
% elev	0.4824	0.4736	0.4198	0.4449
# of pts	60	60	55	48

1st breakpoint

	Glenwood	Nickerson	Legrand	Wolbach
% area	0.22	0.19	0.185	0.26
% elevation	0.0815	0.0596	0.0690	0.0961
# of points	44	38	37	52
H' (scaled)	0.0470	0.0549	0.0508	0.0662

3rd breakpoint

%area	0.75	0.75	0.695	0.805
%elevation	0.6927	0.6164	0.5820	0.6406
# of points	150	150	139	161
H' (scaled)	0.0692	0.0367	0.0524	0.0424

Note: H' (scaled) is the length of the vertical line on the original Lorenz curve

Hoover	0.1732	0.1786	0.1463	0.2124
Hoover Lower	0.0470	0.0549	0.0508	0.0662
Area (weight)	0.445	0.45	0.415	0.565
Hoover Upper	0.0692	0.0367	0.0524	0.0424
Area (weight)	0.555	0.55	0.585	0.435
Gini	0.2325	0.2235	0.1908	0.2683
Full Gini	0.2342	0.2504	0.2027	0.3008

Note: In the table, "r" refers to the transformed elevations after the minimum elevation has been subtracted. "Full Gini" refers to the Gini index using all of the data, and not just the segments created by dividing the Lorenz curve into lower and upper sections.

Example 2

Suppose there are 16 observations of elevation: 8 are at 10 feet, 5 are at 20 feet, and 3 are at 30 feet. Table 3.13 provides a summary.

Table 3.13 Area and elevation data for hypothetical example

	Area			Elevation		
	Freq. (relative)			Total elevation		
Elevation	Rel. freq.	Cum.		Rel. freq.	Freq.	Cumulative
10	8 (0.5)	0.5		80	8/27	0.2964
20	5 (0.3125)	0.8125		100	10/27	0.6667
30	3 (0.1875)	1		90	9/27 = 1/3	1
Total	16			270		

Calculations are as follows:

Hoover index: $0.5 - 0.2964 = 0.2036$; 50% of the area has 29.64% of the elevation. 20.4% of the total elevation would have to be reassigned from high points to low points to flatten the landscape.

Gini coefficient: 0.2388 > 0.2036.

Area *beneath* curve is the sum of a triangle and two trapezoids:
    Triangle: 0.5(0.2964)/2 = 0.0741
    Trapezoid 1: (0.8125 − 0.5)(0.2964 + 0.6667)/2 = 0.1505
    Trapezoid 2: (1 − 0.8125)(1 + 0.6667)/2 = 0.1562

Total area = 0.0741 + 0.1505 + 0.1562 = 0.3808
Gini = (0.5 − 0.3808)/0.5 = 0.1192 × 2 = 0.2384
    23.84% of the mean elevation (0.2384 × 16.875 = 4.02) is the expected
exchange in elevation that would be required to equalize the elevation at two
randomly chosen points.

Mean elevation = [(10 × 8) + (20 × 5) + (30 × 3) = 270]/16 = 16.875
Hyspometric integral = (16.875 − 10)/(30 − 10) = 0.34375
    34.4% of the land mass remains

3.5.2 Surface Metrology

Another active area of application for measures of inequality is that of surface metrology. Metrology is the science of measurement, and surface metrology is concerned with the measurement of small-scale features on surfaces of materials. Surface roughness is one such measure – based on height deviations from a base surface.

There are international standards for such measurements – known as ISO 25178 (International Organization for Standardization). Commencing in 2010, these standards are aimed at the specification and measurement of surface texture measurements. Some parameters are simple and straightforward. For example, Sa is used to designate the arithmetic average absolute deviation from mean; Sq denotes the root mean square of height measurements; Ssk is skewness, and Sku is kurtosis. Other measures are more sophisticated. For example, there is an autocorrelation function defined at each point. The Sal parameter measures the distance to a place on the surface that is significantly different (e.g., where the autocorrelation has declined to 0.2) (Blateyron, 2013).

The **material ratio curve** (see, e.g., Figure 3.4) is the cumulative distribution associated with the density of points that are at a given height. It is also known as the Abbott–Firestone curve (Abbott and Firestone, 1933). It is plotted with height on the

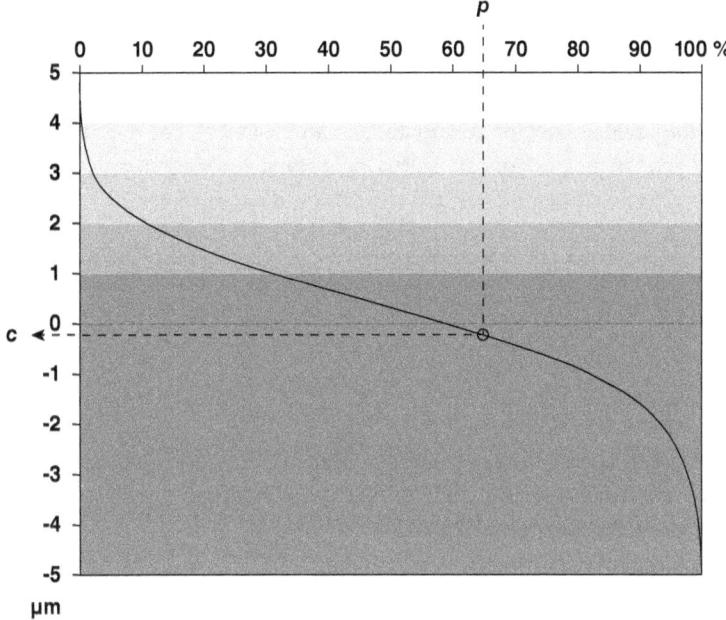

Figure 3.4   Material ratio curve

Source: Blateyron (2013). Reprinted with permission of Springer Nature.

vertical axis, and the cumulative percentage of observations on the horizontal axis. It is thus seen to be directly analogous to the hypsometric curve.

The **areal material ratio**, $Smr(c)$, is the ratio of material p to total material, at height c. This is simply a point on the material ratio curve corresponding to height c.

The **inverse areal material ratio**, $Smc(p)$, gives the height c at which the material ratio is at p.

The **material volume**, $Vm(mr)$, is the volume of material for material ratio mr. For $mr = 100\%$, Vm is a maximum and equivalent to the hypsometric integral.

The use of unitless combinations, analogous to, for example, the coefficient of variation, is not frequent in this subfield, and it would be interesting to explore whether measures such as the Hoover index, Gini coefficient, and hypsometric integral could prove insightful.

As we have seen, there is a wide scope of application for measures of roughness, variability, and inequality. The language used in various subfields is often quite different, and although the methods often share much similarity, there seems to be potential in cross-fertilization, with the measures used in one area possibly being of interest and use in other areas.

Exercises

1. For the following data, find the mean center, and carry out the first two iterations for finding the median center (using Euclidean distance; recall the Pythagorean theorem and the method for calculating the distance between two points). Also find Bachi's standard distance using the result for the mean center.

x	y	w
3	4	100
4	5	200
1	1	100

2. Find the mean center of population for Chicago, Portland (Maine), and Atlanta; use the respective latitude and longitude coordinates (41.878°N, 87.638°W), (43.659°N, 70.257°W), and (33.749°N, 84.388°W). The 2017 populations for Chicago, Portland, and Atlanta were 2,716,000, 66,882, and 486,290, respectively. Use (a) the three-dimensional approach using spherical geometry, and (b) the census approach that uses the Sanson–Flamsteed projection.

Further reading

Centers of population are discussed by Barmore (1993), among many others. More recent references include Abofadel and Austin (2006) and Plane and Rogerson (2015), and the references therein. McCollum (2011) is a good reference for geographic centers. Pike and Wilson (1971) is a classic reference for the hypsometric integral.

Online resources

Visit **https://study.sagepub.com/rogersonspatialstatistics** for resources which supplement the material presented in this chapter. Students can find datasets and additional exercises, which can be used to practice the techniques covered in this chapter, as well as a chapter introduction video and video demonstration of calculating geographic centers. The topic-software chart also describes the software packages that can be used to implement key techniques.

Lecturers and instructors can find exercises and conceptual prompts to facilitate classroom discussion and practice.

4

Statistical Inference and Spatial Patterns

4.1 Types of Data

We'll be looking at two general types of data. One type is point data, where the location coordinates of events are available. These may, for example, be in the form of x- and y-coordinates or latitude and longitude coordinates.

We'll also look at many examples that make use of areal data. Here a region is divided into a number of subregions. Typically for each subregion there is information on the observed number of cases, as well as the number of cases that would be expected, if the null hypothesis (e.g., of spatial randomness) was true. We'll want to compare the observed and expected values, with the intent of finding regions where the observations differ significantly from our expectations.

4.2 Characteristics of Randomness

A random spatial pattern is characterized by two properties. One is that random patterns have mean intensities that do not vary over space. The expected numbers of events in different subregions of equal area are identical. This is known as a first-order property of a point process. We will eventually modify this for our null hypotheses since in practice we will expect *different* numbers of events in subregions of equal size (e.g., since more people may live in one subregion than in another subregion of the same size).

Random point patterns are also characterized by *independence* of point locations. This implies that where one point is located has nothing to do with where other points are located. The covariance of the number of points in one subregion with the number of points in another subregion is expected to be zero. This lack of dependence that characterizes random patterns is known as a second-order property. First-order properties pertain to the mean or a measure of central tendency; second-order properties pertain to the variance or covariance.

4.3 Classes of Questions Regarding Spatial Patterns

We will address three separate types of questions about spatial randomness. Based upon the discussion of Besag and Newell (1991), tests may be (a) global or general tests, (b) local or focused tests, or (c) tests for the detection of clustering (scan-type tests).

4.3.1 Global or General Tests

These tests result in a single observed statistic that characterizes an observed map. We then compare that statistic with a critical value to decide whether the null hypothesis of spatial randomness should be rejected. No information is provided regarding details of possible clusters, including their sizes, shapes, and locations.

4.3.2 Focused or Local Tests

These tests are designed to test the null hypothesis of randomness around a prespecified point – typically against an alternative that there is a clustering of raised risk around the point. The location of interest may be, for example, a source of pollution that is suspected of being associated with a health risk.

4.3.3 Tests for the Detection of Clustering – Scan Tests

Scan tests may be thought of as a series of focused tests. A window is centered on a particular location to see whether there is significant clustering at that location – and then the window is moved across the map to see if there is clustering around *any* location. The challenge here is to adjust properly for all of the multiple testing. With $\alpha = 0.05$, we expect to make a Type I error, on average, once every 20 tests. Thus if we scan a map and evaluate 200 potential sites of clustering, we shouldn't be surprised to see about 10 "significant" clusters – even if the null hypothesis is true. These "clusters" have appeared by chance alone. In our discussion of this type of test, we will consider carefully the issue of multiple testing.

As we turn our interest now to the study of spatial pattern, we are interested in the null hypothesis that a set of observed points comes from a random pattern. Our goal is to determine the *p*-value associated with this hypothesis. How likely would our observed pattern be, if the null hypothesis were true? Possible alternative hypotheses are that the pattern is more clustered than random, or that it is more dispersed than random. Often in spatial applications we use the one-sided alternative hypothesis that the pattern is clustered.

There are several methodological approaches to answering questions like these. One approach is to use an exact approach that makes use of the probability distribution associated with a test statistic. Another is to use an approximation to that exact distribution – often, for example, the central limit theorem is invoked and a normal distribution is assumed for the test statistic, so that a simple *z*-test may be employed. A third approach is to simulate the null hypothesis; if the distribution of the test statistic under the null hypothesis is unknown, and normal approximations are not possible and/or accurate, then it is common and reasonable to simulate the null hypothesis to obtain the probability distribution of the test statistic. This simulated distribution can then be used to obtain critical values for comparison with the observed statistic.

4.4 Circular Study Areas

To begin, consider the case where there is a circular study area, and we are interested in whether points are clustered around the center of the area. The null hypothesis is that the distances from the center to the observed points are consistent with what is expected in a random pattern. The alternative hypothesis might be that the points are more clustered around the center than would be expected by chance alone.

Before proceeding, we note that this is a fairly specialized test that is not useful in many practical circumstances. If we had a map of disease or crime cases, this would be a relevant question *if* the spatial distribution of the underlying population were uniform. In practice of course, population density varies over space. Therefore, there might be clustering of cases that is attributable entirely to the distribution of population. For instance, cases might cluster near the center not because there is raised risk in the vicinity, but rather because many people live there.

For points distributed randomly in a circle, the expected distance to a point from the center is E[d] = 2R/3, where R is the radius of the circle. (Note that the expected distance is more than half of the radius; this is because there is more area near the periphery of the circle than there is near the center.) The variance of the distance from the center to a point is equal to $R^2/18$. This allows us to carry out a z-test for n observed distances, d_i:

$$z = \frac{(\sum_{i=1}^{n} d_i)/n - 2R/3}{\sqrt{R^2/(18n)}}$$
(4.1)

This value of z, based upon the observed data, may be compared with critical values of z. For example, for the one-sided alternative that points are more clustered around the center than in a random pattern, with a Type I error probability of $\alpha = 0.05$, we would use $z_{crit} = -1.644$. The critical value here is negative because we expect the numerator of the z-statistic to be negative, since the observed distance will likely be less than what is expected in a random pattern.

This of course relies upon the assumption that the distribution of mean distances is normally distributed. While means have normal distributions when the sample size is large (regardless of the distribution of the underlying variable), this is not necessarily the case when the sample size is small. This z-test should therefore be regarded as an approximation. We will return to the questions of (a) the formulation of an exact test that does not rely upon the assumption of normality, and (b) how good (or poor) the approximation above actually is.

An alternative test uses the squared distances. In a circle the expected squared distance of a point from the center is equal to $R^2/2$, and the standard deviation of the squared distances is equal to $R^2/\sqrt{12}$. Thus we might use the z-test:

$$z = \frac{\frac{\sum_{i=1}^{n} d_i^2}{n} - \frac{R^2}{2}}{\frac{R^2}{\sqrt{12n}}}$$
(4.2)

where n appears in the denominator of the denominator. The standard deviation of the mean squared distance is smaller for larger sample sizes. Recall that the standard deviation of a mean is equal to $\sigma/\sqrt{n}$.

These z-tests are approximate tests – that is, if the null hypothesis is true, the test statistic will have, only approximately, a normal distribution. The approximation is better as the sample size gets larger. For small samples, though, we have not evaluated how accurate it is – we may reject true null hypotheses either too often, or not often enough.

4.4.1 Simulation Approach

There are two possible alternatives we can consider. The first is simulation. We can simu-late the null hypothesis, randomly placing n points in a circle of radius 1. To do this, we can choose a random number, u, from a uniform distribution on the interval $(0,1)$, and set it equal to the cumulative distribution function of distances (see the discussion of simulation in Appendix A). The probability that a distance is less than r is equal to the area of the circle with radius r, divided by the size of the study area. Thus

$$u = F(r) = \Pr(distance < r) = \frac{\pi r^2}{\pi R^2} = \frac{r^2}{R^2} \tag{4.3}$$

Thus a randomly chosen distance from the center can be chosen by solving (4.3) for r:

$$r = \sqrt{u}R \tag{4.4}$$

To place the point at a precise location, we also need to consider where to put it, along the circumference of the circle that surrounds the center at a distance r. We designate this position along the circumference of the circle (with radius r) with the notation θ, where θ represents an angle between 0° and 360°. For this, we choose another uniform random number between 0° and 1, and multiply by 360°. This gives us a (r,θ) coordinate pair within the circle.

So one way to test the null hypothesis of a random pattern is to choose n points in the way just described. Then find the mean distance of these points from the center of the circle. Then this is repeated many times, say 1,000 times. These 1,000 mean distances can be ordered from lowest to highest, and we can make a histogram using them. We want to know whether our observed mean distance from the center is an unusual one, when compared with the 1,000 simulated values. If our value is very small, say lower than the 50th lowest number on our list of 1,000 (associated with a one-sided alternative hypoth-esis, and a Type I error probability of $\alpha = 0.05$; since 50/1,000 = 0.05), then we would reject the null hypothesis and conclude that the pattern was more clustered than ran-dom. The 50th lowest number on the list of simulated means serves as the critical value.

It is important to recognize that simulation will result in different critical values every time we repeat the process. In fact, it is good practice to do exactly that – repeat the (say 1,000) simulations three or four times. If you get the roughly the same critical value each time, this provides some assurance that the number of simulations (here equal to 1,000) is sufficient. If the critical values differ significantly, this suggests that the number of simulations should be increased (e.g., from 1,000 to 5,000 or 10,000).

The simulation approach is particularly useful when the distribution of the test statis-tic is either unknown or its computation (and in particular computation of critical values and percentiles) is difficult. If the distribution of the test statistic is known and reasona-ble from a computational perspective, then it should be used instead of simulation.

It would be interesting to compare the results of this approach with the z-test previously described, for different values of n. For large values of n, we would find that the two approaches would give us similar results. But how would the results begin to diverge, as we compared the two tests for smaller and smaller values of n? In particular, we would find that with the simulation approach, we would reject null hypotheses the correct proportion of times. But the z-test would either reject too often or not often enough, and in principle, the critical value of z (of -1.645, for a one-sided test with $\alpha = 0.05$) should be adjusted to give the desired level of a Type I error. What should it be adjusted to? It should be adjusted to the 50th lowest number on our list of simulated z-scores. This will likely be somewhere near -1.645, but will not be precisely equal to it. To address these questions, a series of simulations was carried out. The simulations were performed for two cases – one where $n = 10$, and one where $n = 20$. For each of these two choices, n points were located in a circular study area, and n points were located within a square study area. And in each of *these* cases, the distances and squared distances from the center to each of the n points was noted. Finally, the z-scores were calculated and this was repeated for each case 100,000 times. The z-scores were ordered, and the 5th and 95th percentiles were recorded. They appear in Table 4.1.

Table 4.1   The 5th and 95th percentiles of the distribution of distance and squared distance, for circular and square study areas

	Distance percentile		Squared distance percentile	
Circle	5th	95th	5th	95th
$n = 10$	−1.653	1.644	−1.698	1.595
% of z-score < −1.645	(0.0509)		(0.0555)	
$n = 20$	−1.636	1.647	−1.675	1.605
% of z-score < −1.645	(0.0492)		(0.0534)	
Square				
$n = 10$	−1.694	1.607	−1.602	1.682
% of z-score < −1.645	(0.0543)		(0.0453)	
$n = 20$	−1.678	1.615	−1.613	1.672
% of z-score < −1.645	(0.0524)		(0.0464)	

All of the results are not too far from the usual critical values of -1.645 and 1.645 that would be used for a normal distribution. This is to be expected since means have normal distributions for large sample sizes. Entries in the table are a little farther from these tabled values when the sample size is smaller ($n = 10$). For the 5th percentile, the percentage of simulations having a z-score less than -1.645 is given in parentheses. These correspond to the fraction of Type I errors that would be expected, *if* the usual critical value of -1.645 were used. These are near the nominal value of 0.05. It is interesting to note that for circular study areas, the use of the normal approximation with distance appears to be slightly more accurate than with the use of squared distance. For square study areas, there is less of a difference between the two choices. It is also interesting to note that for square regions, the use of squared distance is "conservative" (i.e., the values in parentheses are less than the nominal value of $\alpha = 0.05$, indicating that we would reject null hypotheses a little less often than we should), while the use of distance in the calculation of z is "liberal" (the null hypothesis is rejected a little too frequently). For circular study areas, the use of squared distance seems to be a little more liberal than when distance is used.

Again, we emphasize here that a major aim of the above is to introduce and illustrate ideas and concepts; less important is the introduction of a specific method that will actually turn out to be used frequently in practice.

4.4.2 An Exact Approach

A more exact test is possible if we consider the exact distribution of the mean of squared distances. We have already seen that the distance of a random point from the center has a distribution that is equivalent to whatever distribution the square root of a uniform random variable has. Thus the squared distance of a random point from the center will follow a uniform distribution. The lowest this variable could be is 0, if the point chosen is right at the center of the circle, and the largest it could be is R^2, which would happen if the point was on the periphery. Under the null hypothesis of spatial randomness, the mean of this random variable, as we have already seen, is $R^2/2$.

What we want to know here is the distribution of the mean of squared distances. We'd like to compare our observed mean of the squared distances with this distribution to assess whether our observed value is unlikely to have occurred by chance, if the null hypothesis is true. This entails adding up a set of n squared distances, and then dividing by n. This is sometimes called the null distribution – it is the distribution we would expect if the null hypothesis were true. Equivalently, we can ask what the distribution of a sum of n uniform random variables (on the interval between 0 and 1) looks like (since the mean only differs from the sum by a scalar factor of $1/n$). It turns out that this is a distribution that has been derived – it is known as the Irwin–Hall distribution. Its cumulative distribution function is given by

$$F(x) = \frac{\sum_{k=0}^{\lfloor x \rfloor}(-1)^k \binom{n}{k}(x-k)^n}{n!} \tag{4.5}$$

where the notation $\lfloor x \rfloor$ refers to the greatest integer that is less than or equal to x. The quantity k is simply an index and has no other interpretation. One could test the null hypothesis by comparing the observed mean distance with the 5th or 95th percentile of this distribution. For example, with $n = 5$ points, we would expect the sum of the squared distances to be less than 1.5 a little more than 6% of the time, e, when $R = 1$:

$$F(1.5) = \frac{\binom{5}{0}(1.5-0)^5 - \binom{5}{1}(1.5-1)^5}{5!} = 0.062 \qquad (4.6)$$

In this example, $[1.5] = 1$ and thus k can be either 0 or 1. Therefore, there are two items in the numerator. The reader may wish to verify that a value of about $x = 1.43$ will lead to a value for $F(x)$ of approximately 0.05. Thus a sum of squared distances of 1.43 (or equivalently, an average squared distance of $1.43/5 = 0.286$) could be used as a critical value. An observed value less than this would not be found often if the pattern were random, and hence the null hypothesis would be rejected in favor of the alternative that the points were more clustered around the center than would be the case in a random pattern.

This is not widely used within the field of spatial statistics – in fact, it may *never* have been used in this context! The main point here, though, is to lay out different ways to answer questions. Here we have described three ways – one is simulation, and the others are analytical (with the z-test being an approximate analytical approach, and where this last approach constitutes an exact analytical approach). Each of these approaches has its merits. The simulation approach is sometimes the only reasonable approach that can be used – this is the case when the analytical solution is too complex. A drawback of the simulation approach is that you get a different answer every time you use it. The exact analytical approach is in some sense the most desirable – an exact answer (in this case, a precise p-value) is provided, but often the path to the answer is complex enough that the idea is abandoned in favor of an approximate test, such as the z-test described above.

4.5 Square Study Areas

There are similar tests that can be used if the study area is a square. In a square, the expected distance from a randomly chosen point to the center is equal to $(s/6)(\sqrt{2} + \ln(1+\sqrt{2}))$, where s is equal to the length of a side of the square. This is approximately equal to $0.383s$. The variance of these distances is $s^2/6 - (0.383s)^2$, which is approximately equal to $0.02s^2$. A z-test based on this is

$$z = \frac{\bar{d} - 0.383s}{s\sqrt{0.02/n}} \qquad (4.7)$$

The expected squared distance from a randomly chosen point to the center of a square is equal to $s^2/6$, and the standard deviation of the distances of randomly chosen points to the center is equal to $s^2/\sqrt{90}$. A corresponding z-test is

$$Z = \frac{\dfrac{\sum_{i=1}^{n} d_i^2}{n} - \dfrac{s^2}{6}}{\dfrac{s^2}{\sqrt{90n}}} \qquad (4.8)$$

4.5.1 Simulation

The distribution of mean distances could be found by simulation. To simulate points within a square, we can simply choose two random, uniformly distributed numbers between 0 and 1 – one serves as the x-coordinate and one as the y-coordinate. If the study area is not scaled to have maximum coordinates of 1, then it will be necessary to multiply each of the two numbers by a scaling factor to ensure that all values between 0 and s are equally likely, where s is the length of a side of the square.

In the case of a square, finding the exact distribution of the mean distance of n points from the center is analytically difficult. But Weissman (2017) shows that for the distribution of $x = a^2 + b^2$, where a and b are uniform variables on the interval $(0,1)$, the probability density function is

$$\begin{aligned} f(x) &= \pi/4; \ 0 < x < 1 \\ &= \sin^{-1}\left(x^{-1/2}\right) - \pi/4; \ 1 < x < 2 \end{aligned} \qquad (4.9)$$

and the cumulative distribution function is

$$\begin{aligned} & \pi x/4; \ 0 < x < 1 \\ & (x-1)^{1/2} + \sin^{-1}\left(x^{-1/2}\right) - \pi x/4; \ 1 < x < 2 \end{aligned} \qquad (4.10)$$

This is the distribution of the squared distance from the origin to a point randomly chosen in the unit square.

Now, if we rescale our square study area to have ranges along x- and y-axes from -1 to $+1$, our fixed point is at the middle of this relabeled area (at $(0,0)$), and there are four quadrants that a random point could fall into. The point will fall randomly into any particular quadrant with probability one-fourth, and the probability that it is a given (squared) distance from the center is given by the probability density function above. Note that the largest value of the squared distance is 2 – this would occur when x and y both have absolute values of 1.

---------------------------------(**Example**)---------------------------------

Suppose for a square study area with range $(-1, +1)$ for x- and y-coordinates, $n = 5$, and the points are located at the following (x, y) locations: $(0.5, 0.7)$, $(0.2, 0.2)$, $(0.1, 0.2)$, $(0.4, 0.3)$, and $(0.4, 0.4)$. (Note here that for simplicity, and without loss of generality, all of the x- and y-coordinates are positive. The example could just as easily be constructed with negative x- and/or y-coordinates, and this would not change the answer.) Test the null hypothesis that the points are distributed at random distances from the middle of the square study area.

Solution

First note that the locations of four of the five points are quite close to the center of the square, and hence we will not be surprised if we reject the null hypothesis. We will test this hypothesis by testing the null hypothesis that the empirical distribution function of observed squared distances is no different from the cumulative distribution function given above in Equation 4.10. The respective squared distances of the five points from the middle of the square are $(0.5^2 + 0.7^2 = 0.74, 0.2^2+0.2^2 = 0.08, 0.05, 0.25$, and $0.32)$. If the null hypothesis is true, we expect the cumulative distribution to follow Equation 4.10. The Cramér–von Mises test is a test that is relatively powerful in this situation where we are interested in rejecting alternatives where the mean distance of the points is closer to the center than expected (see, e.g., Stephens, 1974). The test statistic is denoted W^2, and it is based upon the sum of squared differences between the observed and expected cumulative distributions. It is calculated as

$$W^2 = \frac{1}{12n} + \sum_{i=1}^{n}\left(\frac{2i-1}{2n} - F(x_i)\right)^2 \tag{4.11}$$

where the observations $\{x_1, ..., x_n\}$ have been arranged in order, from lowest to highest, and where $F(x_i)$ is the probability of obtaining a value from the distribution that is less than x_i (and in this example that is obtained from Equation 4.10).

Here, we have

$$W^2 = \frac{1}{12(5)} + \left(0.1 - \frac{0.05\pi}{4}\right)^2 + \left(0.3 - \frac{0.08\pi}{4}\right)^2 + \left(0.5 - \frac{0.25\pi}{4}\right)^2 +$$
$$\left(0.7 - \frac{0.32\pi}{4}\right)^2 + \left(0.9 - \frac{0.64\pi}{4}\right)^2 = 0.472 \tag{4.12}$$

To adjust for small sample sizes, Stephens (1974: 732) suggests the modified value

$$\tilde{W}^2 = (W^2 - 0.4/n + 0.6/n^2)(1+1/n) \tag{4.13}$$

The result, $\tilde{W}^2 = 0.499$, exceeds the critical value of 0.461 (using a Type I error probability of $\alpha = 0.05$ and the table of critical values given in the paper by Stephens) and so the null hypothesis is rejected. The points are unlikely to have occurred randomly with respect to their distances from the middle of the square.

Solution by Simulation

Five points were chosen at random within the unit square; for each, the squared distance from the origin was calculated. With 100,000 such simulations, the sum of squared distances was less than 1.32 only 1% of the time. This can be used as a critical value ($\alpha = 0.01$) against which the observed value of $0.05 + 0.08 + 0.25 + 0.32 + 0.74 = 1.44$ may be compared. The observed value is just greater than this critical value and hence the null hypothesis is rejected – these points are closer to the origin than would be expected, had the points been distributed randomly.

Simulation for Non-Uniform Population Distributions

We can also look at the case of nonuniform population distributions. Up until this point, we have implicitly assumed a homogeneous population. However, it is easily recognized that in almost all practical cases, we will not *expect* the distribution of points to be random – disease cases and crime cases, for example, are clustered because the population itself is clustered. A far more interesting hypothesis than that of a random spatial pattern with homogeneous population is a random spatial pattern, accounting for the non-uniform distribution of population. This case will be examined in more detail later on, but for now, let us examine how we might simulate case locations for a non-uniform population.

A hypothetical population is shown in Figure 4.1: 300 people live in the northern part of the study area (A), 200 live in the southwest quarter (B), and 100 live in the southeast quarter (C). Suppose that we want to locate points at random in this study area, making sure that the distribution of points reflects the distribution of population. The first step is to choose which neighborhood to locate a point in. To do so, choose a random number between 0 and 1. If it is between 0 and 0.5, the next point will be located in neighborhood A; if it is between 0.5 and 0.8333, it will be located in neighborhood B; and if it is greater than 0.83333, it will be located in neighborhood C. Thus we divide the number line between 0 and 1 into three parts, one for each neighborhood, with each part of the line proportional to population (rather than area). Thus neighborhood A will receive on average half of the points, B will receive on average one-third ($0.8333 - 0.5 = 0.3333$) of the points, and C will receive about one-sixth ($1 - 0.8333 = 0.1667$) of the points. Once we have decided which neighborhood the point is going to be in, we can simply use the uniform distribution to choose x- and y-coordinates for that neighborhood.

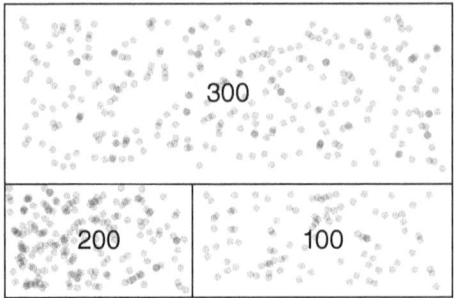

Figure 4.1   Distribution of hypothetical population

One reason for carrying out such simulations is to compare an observed map pattern with the simulations, which reflect how we would expect a pattern to look if the disease or crime were distributed in proportion to population size.

Simulation for Irregular Study Areas

When the boundary of the region is not regular (e.g., a circle or square), random locations within the region can be chosen as follows. Choose a random number from a uniform distribution that ranges from x_{min} to x_{max} where x_{min} is the smallest x-value that is observed and x_{max} is the largest. Another uniform random variable is chosen from the range (y_{min}, y_{max}). If the chosen coordinate pair lies within the region, we have our random location. If it does not lie within the region, we simply discard it and try again.

4.6 Global Tests: History and Perspective

Historically, global tests for the assessment of spatial pattern emerged from the literature in ecology. One such test is the nearest neighbor statistic; although its frequency of application has now been superseded by other global statistics, we include it here because of its historical significance in both ecology and the development of the quantitative revolution in geography.

4.6.1 Nearest Neighbor Statistic

Clark and Evans (1954) developed nearest neighbor analysis to analyze the spatial distribution of plant species. They developed a method for comparing the observed average distance between points (mapped in a two-dimensional plane) and their nearest neighbors with the distance that would be expected between nearest neighbors in a random pattern.

We begin by defining R_0 to be the observed average distance between points and their nearest neighbors. Let R_e be the expected distance between points and their nearest neighbors when points are distributed randomly. Intuitively, if R_0 is small relative to R_e, the pattern will be clustered; if R_0 is large relative to R_e, the pattern will be more dispersed than random.

R_0 may be calculated as $\sum_{i=1}^{n} d_i / n$ where n is the number of points in the study area, and where d_i is the distance from point i to its nearest neighbor. Note that nearest neighbors may be reflexive – that is, they may be nearest neighbors of each other. For example, in Figure 4.2, there are 6 points. Point A is 3 units away from its nearest neighbor. Points B and C are reflexive nearest neighbors; they are each 2 units away from their nearest neighbors. Point D is 4 units away from its nearest neighbor and, finally, points E and F are reflexive nearest neighbors, 1 unit away from each other. The average nearest neighbour distance is $(3 + 2 + 2 + 4 + 1 + 1)/6 = 2.17$ units.

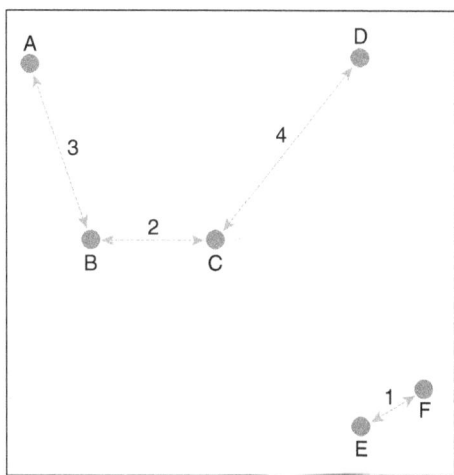

Figure 4.2 Hypothetical data for calculation of nearest neighbor statistic

R_e is calculated as one over twice the square root of the density of points:

$$R_e = \frac{1}{2\sqrt{\rho}}$$ (4.14)

where ρ is the density of points and is equal to the number of points (n) divided by the size of the study area (A). In Figure 4.2, the study region has an area of 36 units, and so the density is equal to $6/36 = 0.167$. Hence the expected distance between nearest neighbors is $1/(2\sqrt{(6/36)}) = 1/0.816 = 1.22$. Our pattern is more spread out than would be expected since the observed average distance between neighbors (2.17) is greater than expected.

The nearest neighbor statistic, R, is sometimes defined as the ratio between the observed and expected values:

$$R = \frac{R_o}{R_e} = \frac{\bar{d}}{1/\left(2\sqrt{\rho}\right)} = 2\bar{d}\sqrt{\rho}$$ (4.15)

R varies from 0 (a value obtained when all points are in one location), to a theoretical maximum of about 2.14, for a perfectly uniform or systematic pattern of points maximally spread out on an infinitely large two-dimensional plane (in practice, for bounded regions, the ratio can exceed 2.14). A value of $R = 1$ indicates a random pattern, since the observed mean distance between neighbors is equal to that expected in a random pattern. In the example in Figure 4.2, $R = 2(2.17)\sqrt{6/36} = 1.78$.

It is also known that if we examined many random patterns, we would find that the variance of the observed mean nearest neighbor statistic, R_0, is

$$V[R_0] = \frac{0.0683}{n\rho}$$ (4.16)

where n is the number of points. Thus, we can form a z-test, to test the null hypothesis that the pattern is random with the now familiar process of starting with the statistic (R_0), subtracting its expected value, and dividing by its standard deviation to obtain a z-score:

$$z = \frac{R_0 - R_e}{\sqrt{V[R_0]}}$$ (4.17)

The quantity z has, approximately, a normal distribution with mean 0 and variance 1, and hence tables of the standard normal distribution may be used to assess significance. A value of $z > 1.96$ implies that the pattern has significant uniformity, and a value of $z < -1.96$ implies that there is a significant tendency toward clustering. For the pattern in Figure 4.2, $z = (2.17-1.22)/\sqrt{(0.0683/(6 \times 6/36))} = 3.6$, and this implies that the null hypothesis of spatial randomness can be rejected in favour of the alternative that the points are more spread out than would be expected.

The strength of this approach lies in its ease of both calculation and comprehension. Several cautions should be noted in the interpretation of the nearest neighbor statistic. The statistic, and its associated test of significance, may be affected by the shape of the region. Long, narrow, rectangular shapes may have relatively low values of R simply because of the constraints imposed by the region's shape. Points in long, narrow rectangles are *necessarily* close to one another.

The location of points relative to the boundary of the study region can also make a difference to the analysis. Points near the boundary may actually have their nearest neighbor just outside of the boundary. The average distance to the nearest neighbor is therefore overestimated, and the z-statistic is higher than it would be, had edge effects been accounted for. Thus when clustering exists, we may fail to reject a test of the null hypothesis against the alternative of clustering, thereby committing a Type II error.

One solution to the boundary problem is to place a buffer area around the study area. The nearest neighbors are found for all points within the study area (but not for the points in the buffer area). Points inside of the study area (such as point A in Figure 4.3) may have nearest neighbors that fall into the buffer area, and these distances should be used in the analysis.

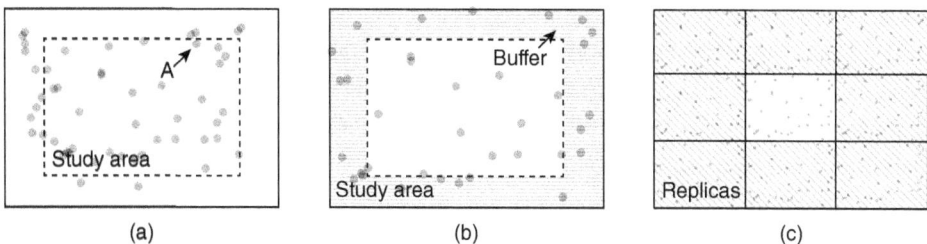

(a)   (b)   (c)

Figure 4.3a, b, c Consideration of edge effects

If information on the location of points just outside of the study area cannot be obtained, then one can create a similar buffer region, but with it now lying *inside* of the study area. This has the effect of creating a new study area that is effectively smaller than the original one. While having the benefit of accounting for edge effects, it has the necessary disadvantage of using less information due to the smaller size of the modified study area (Figure 4.3b).

An alternative is to attach replicas of the study region and its observed points to its sides (Figure 4.3c). A variation of this would be to attach a buffer area, and randomly scatter points at the observed density throughout the buffer region.

Yet another approach to addressing edge effects is to use a Monte Carlo simulation. With the Monte Carlo approach, the null hypothesis (in this case, that the point pattern is random) is simulated by choosing the same number of points as is observed, and randomly placing them within the study area. This random placement could be carried out with the aid of software, a calculator, or a table of random numbers. Then the statistic (R) is found for this simulated pattern. This step is then repeated a large number of times to generate many values of R, preferably using a computer. The value of R observed for the actual map is compared with these simulated values to see how unusual it is, to decide if there is enough evidence to reject the null hypothesis. If the observed R is unusual relative to the simulated values (e.g., if a value more extreme than the one observed would occur less than 5% of the time), the null hypothesis is rejected. This Monte Carlo method of simulating the null hypothesis many times, and then comparing the observed result to the simulations, is a very general approach, and can be used in a wide range of situations to test hypotheses.

Another potential difficulty with the nearest neighbor statistic is that, since only nearest neighbor distances are used, clustering is only detected on a relatively small spatial scale. Distances to nearest neighbors may be consistent with the expected distances in a random pattern, but what about distances to second nearest neighbors, third nearest neighbors, and so on? To overcome this limitation, it is possible to extend the approach to second- and higher order nearest neighbors. The expectations and variances for these distances are discussed by Wilson and Kirkby (1975), but they are not often used in practice and discussion of the details of that extension is beyond the scope of this text.

Most importantly, it is often of interest to ask not only whether clustering exists, but whether clustering exists over and above some background factor (such as population). For instance, using the nearest neighbor statistic to determine whether crime is clustered in an urban area is often not too enlightening – the nearest neighbor analysis likely reveals clustering because the population itself tends to be clustered. Nearest neighbor methods are not particularly useful in these situations because they only relate to the spatial location of the points, and do not account for other factors that are already known to influence the spatial distribution of points. Many of the approaches to the study of pattern that are described in Chapters 5, 6, and 7 do not have this limitation.

Illustration: Nearest neighbor distances

For the point pattern in Figure 4.4, there are six point locations (A through F).

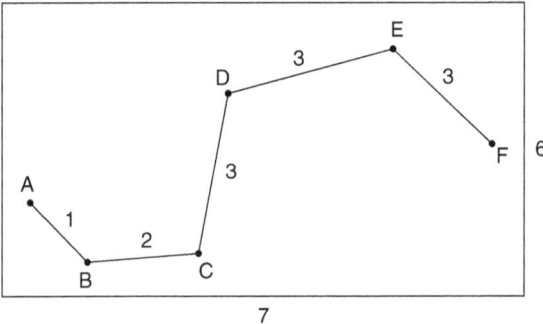

Figure 4.4   Nearest neighbor distances

Distances between points are given along the lines connecting the points. The mean distance between nearest neighbors is $R_0 = (1 + 2 + 3 + 1 + 3 + 3)/6 = 13/6 = 2.167$. The expected mean distance between nearest neighbors in a pattern of six points placed randomly in a study region with area $7 \times 6 = 42$ is

$$R_e = \frac{1}{2\sqrt{\rho}} = \frac{1}{2\sqrt{6/42}} = 1.323 \tag{4.18}$$

The nearest neighbor statistic is $R = 2.167/1.323 = 1.638$, which means that the pattern displays a tendency toward uniformity. To assess significance, we can calculate the z-statistic from $(2.167 - 1.323)/\sqrt{(0.0683/(6/7))} = 2.99$, and this is much greater than the critical value of 1.96, which in turn implies rejection of the null hypothesis of a random pattern. However, we have neglected boundary effects, and these can have a significant effect on the results. As an alternative way to test the null hypothesis, we can randomly choose six points by choosing random x-coordinates in the range (0, 7) and random y-coordinates in the range (0, 6). Then we compute the mean distance from each of the six points to their nearest neighbors, and repeat the whole process many times. Simulating the random placement of six points in the 7×6 study region 10,000 times led to a mean distance between nearest neighbors of 1.62. This is greater than the expected distance of $R_e = 1.323$ noted above. This greater-than-expected distance can be attributed directly to the fact that points near the border of the study region are relatively farther from other points in the study region than they presumably would have been to points just outside of the study region. Ordering the 10,000 mean distances to nearest neighbors from lowest to highest reveals that the 500th highest one is 2.29. This implies that only 5% of the time would we expect a mean nearest neighbor distance greater than 2.29. Our observed distance of 2.167 is less than 2.29, and so we, having accounted for boundary effects through our Monte Carlo simulation, fail to reject the null hypothesis.

In an appendix to their paper, Clark and Evans provide a derivation of the expectation and variance of the distance from a point to its nearest neighbor. The derivation assumes that points are distributed randomly in an infinite plane, where there are no edges and border effects. A summary of the derivation follows (see Appendix A for some background material to assist with following the details).

If the density of points is equal to ρ, then the number of points expected in an area of size A is equal to ρA. The expected number of points in a circle of radius r that surround a point is therefore equal to $\rho \pi r^2$, and this can become the parameter of a Poisson distribution (which describes a random spatial distribution of points). The probability that there are no other points in the circle of radius r that surrounds a point is $\Pr(X = 0)$, and is also equal to the probability that the nearest neighbor distance is greater than r:

$$\Pr(X = 0) = \frac{e^{-\rho \pi r^2} \left(\rho \pi r^2 \right)^0}{0!} = e^{-\rho \pi r^2} = 1 - F(r) \tag{4.19}$$

where $F(r)$ is the cumulative distribution function associated with the probability distribution of nearest neighbor distances. Thus

$$F(r) = \Pr(\text{nearest neighbor distance} < r) = 1 - e^{-\rho \pi r^2} \tag{4.20}$$

The slope of the cumulative distribution function is equal to the probability density function. Taking the derivative of $F(r)$ with respect to r yields the probability density function:

$$f(r) = 2 \rho \pi r e^{-\rho \pi r^2} \tag{4.21}$$

The expected value of the nearest neighbor distance r in a random pattern is

$$E[r] = \int_0^\infty r f(r) dr = 2 \rho \pi \int_0^\infty r^2 e^{-\rho \pi r^2} dr = \frac{1}{2\sqrt{\rho}} \tag{4.22}$$

The variance is found as

$$\sigma^2 = V[r] = E[r^2] - \{E[r]\}^2 = 2\rho\pi \int_0^\infty r^3 e^{-\rho\pi r^2} dr - \left(\frac{1}{2\sqrt{\rho}}\right)^2 = \frac{4-\pi}{4\pi\rho} \approx \frac{0.0683}{\rho} \tag{4.23}$$

Exercises

1. A circular study area with a radius of 10 km has a hazardous waste site at its center. Cases of disease are found at distances of 1.1 km, 2.3 km, 4.1 km, 5 km, and 7.2 km from the center. Use the z-tests in (4.1) and (4.2) to test the null hypothesis that these cases are randomly distributed within the study area.
2. Assume that a square study area with area 324 km² has a hazardous waste site at its center. Cases of disease are found at distances of 1.1 km, 2.3 km, 4.1 km, 5 km, and 7.2 km from the center. Use the z-tests in (4.7) and (4.8) to test the null hypothesis that these cases are randomly distributed within the study area.
3. A study area has an area of 100 km². An investigation of eight arsons over a period of time reveals that the set of distances from each arson to the closest neighboring arson is {1, 1.5, 1.7, 1.8, 1.3, 2.4, 2.1, 0.8}. Use the nearest neighbor test in (4.17) to test the null hypothesis that the arsons are randomly distributed within the study area.

Further reading

Besag and Newell (1991) published a very important and widely cited paper on types of questions pertaining to geographic clustering. The classic reference for the nearest neighbor statistic is Clark and Evans (1954).

Online resources

Visit **https://study.sagepub.com/rogersonspatialstatistics** for resources which supplement the material presented in this chapter. Students can find datasets, which can be used to practice the techniques covered in this chapter, as well as a chapter introduction video and video demonstration of calculating the nearest neighbour statistic.

Lecturers and instructors can find exercises and conceptual prompts to facilitate classroom discussion and practice.

5

Global Tests

Chapter overview

In this chapter, you will learn about:

- Quadrat tests
- Moran's *I* and Geary's *C*
- Tango's statistic
- Spatial version of the chi-square statistic

Global statistics (sometimes called general statistics) give us a single number that assesses map pattern. This allows us to assess whether the pattern is random, clustered, or more spread out than random. These statistics therefore allow us to test null hypotheses about spatial randomness.

5.1 Quadrat Tests

Quadrat tests of spatial randomness were developed in the field of ecology (see, e.g., Gleason, 1920). They are carried out by overlaying a square grid containing m cells on top of a study area containing n points. Counts are made of the number of events or points in each cell. For a random pattern, the probability distribution of counts follows a Poisson distribution. Specifically, for a Poisson distribution, the variance of the number of points per cell is equal to the mean number of points per cell. This suggests that the variance-to-mean ratio (VMR) may be used for a statistical test of randomness. The VMR (also known as the "index of dispersion") may be expressed as

$$\text{VMR} = \frac{\sum_{i=1}^{m}(x_i - \bar{x})^2}{(m-1)\bar{x}} \tag{5.1}$$

where $\bar{x}$ is equal to n/m.

The chi-square statistic associated with the conditional quadrat test is equal to the VMR, multiplied by $(m-1)$:

$$\chi^2 = (m-1)\text{VMR} = \frac{(m-1)\sum_{i=1}^{m}(x_i - \bar{x})^2}{(m-1)\bar{x}} = \frac{\sum_{i=1}^{m}(x_i - \bar{x})^2}{\bar{x}} \tag{5.2}$$

The word "conditional" refers to the fact that the statistic's null distribution is conditional upon there being a total of exactly n points that are randomly distributed across the quadrats.

The statistic may be compared with critical values from a chi-square distribution with $m-1$ degrees of freedom. A low value of the statistic occurs when the variance-to-mean ratio is low. In turn, this occurs when there is little variation in the number of points in each cell. This characterizes situations where there is dispersion – the point distribution is more regular than random. Similarly, the value of the statistic is high when the variance is high relative to the mean – there is spatial clustering, with few points in some cells and many points in others.

Example

For the 5 x 5 set of cells in Figure 5.1, there are 9 cells with no points, 6 cells with one point, 5 cells with two points, 3 cells with three points, and 2 cells with four points. The average number of points per cell is 33/25 = 1.32. The statistic is

$$\frac{9(0-1.32)^2 + 6(1-1.32)^2 + 5(2-1.32)^2 + 3(3-1.32)^2 + 2(4-1.32)^2}{1.32} = 31.39 \tag{5.3}$$

This is less than the critical value of 36.4 for a chi-square variable with 25−1 = 24 degrees of freedom and therefore the null hypothesis is not rejected. The p-value is equal to 0.143 for a one-sided test; about 14% of the time we would expect to obtain a higher value of chi-square than the one observed, when the null hypothesis is true (see, e.g., the chi-square table available at this book's website).

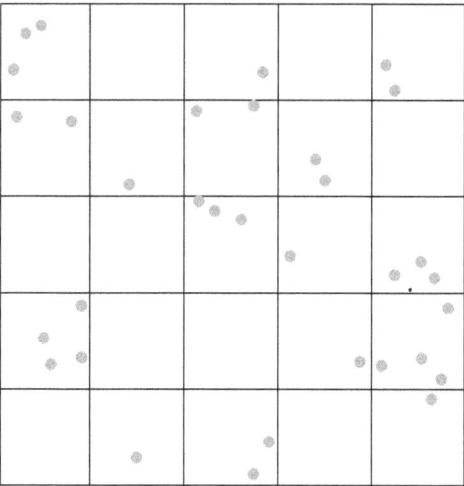

Figure 5.1 Hypothetical point distribution for quadrat analysis

The expected value of the chi-square statistic, when the null hypothesis is true, is equal to $m - 1$. This is intuitive given its definition as $(m - 1)$VMR, since the expected value of VMR is 1, when the null hypothesis is true.

In Chapter 1, we carried out an exercise where we placed 20 points on a map with 25 cells. The mean number of points per cell was $20/25 = 0.8$. We gave each cell a score of 0.8 if there were 0 points in the cell – note now that this is equal to $(0-0.8)^2/0.8$. When there was one point in a cell, the score was 0.05, and now we note that this is equal to $(1-0.8)^2/0.8$. When there are n points in a cell, the score for the cell is $(n-0.8)^2/0.8$. The score for the map was the sum of the scores across all cells or quadrats. The score for the map was the chi-square statistic for the conditional quadrat test. Any score less than $m-1 = 24$ indicated a tendency toward a pattern that is more regular than random, and a score greater than that indicated a tendency toward clustering.

5.1.1 Generalization to Chi-square Goodness-of-Fit Tests

In the previous section, the expected number of points or events falling into each quadrat was the same (and equal to the mean, $\bar{x}$). This is a clear limitation in applications, since the expected number of points in a subarea might be a function of other factors. For example, if the points represented cases of disease, we might *expect* more cases in quadrats where the population was higher, or where the size of the older population was higher. It is possible to generalize the quadrat test to allow for differing expectations in each cell:

$$\chi^2 = \sum_{i=1}^{m} \frac{(x_i - e_i)^2}{e_i} \tag{5.4}$$

where e_i represents the expected number of events in cell i. Furthermore, there is no reason that the quadrats have to be square cells of equal size; they can instead be subregions of any size or shape. This generalized form of the test is simply the well-known chi-square goodness-of-fit test.

5.1.2 Minimal Expectations

A longstanding and widely applied rule-of-thumb is that the expected values for each cell or subregion should be at least five. Otherwise, the distribution of the statistic under the null hypothesis will begin to diverge from a chi-square distribution. However, this is recognized to be conservative, and in practice expectations as low as one do not lead to significant inaccuracies.

5.1.3 Scale

This test will have the highest statistical power (i.e., it will be most effective at rejecting false null hypotheses) when the size and location of any cluster of points match the size and location of the quadrat. If quadrats are small in comparison to the size of the cluster, the effect of the cluster on the statistic will be muted by the fact that the cluster is spread over several cells. Similarly, if there is a small cluster within a large quadrat, the statistic may not capture the cluster. The statistic is based only on the *number* of points within a quadrat, and doesn't account for the fact that the points might be clustered *within* it. Several studies have therefore attempted to specify the "optimal" size of quadrats (see, e.g., Curtis and McIntosh, 1950). However, if the intent is to find clusters, these suggestions are only good if the cluster happens to be about the size of the resulting quadrat. These guidelines have to be balanced with the point made above that the best quadrat size is the one that is identical to the size of the cluster (this is sometimes referred to as the "matched-filter" theorem). Unfortunately one does *not* know the size of the cluster – after all, that is why the test is carried out! But if there is some a priori hypothesis about the possible cluster size, it is not a bad idea to choose the quadrat size as that cluster size.

   It is also possible that the pattern might be very dispersed (uniform and regular are synonyms with "dispersed" in this context) in part of the study area, and very clustered in another part of the study area. A single global statistic (e.g., the global chi-square statistic) may in this case fail to reject the null hypothesis of spatial randomness.

Although the pattern is far from random, the non-significant statistic reflects an average of a partially dispersed and partially clustered pattern.

This is all related to the point that, in a sense, the chi-square goodness-of-fit test is not really a *spatial* test. The statistic compares observations and expectations for each quadrat, but does not recognize *where* the significant deviations are. These deviations could be from cells that are adjacent to one another (implying the existence of a large cluster), or the deviations could be from cells that are widely separated. In Section 5.5 we will explore a spatial version of the chi-square goodness-of-fit test that accounts for the spatial pattern of deviations between observations and expectations.

5.1.4 Multiple Testing of Different Grid Sizes

Given that results will change with different grid sizes (an instance of the well-known modifiable areal unit problem, or MAUP), it is natural to try different grid sizes. However, if different grid sizes are explored for a given dataset, the problem of multiple testing must be accounted for. The more tests we try, the more likely it is that at least one of those tests is found significant by chance alone, even when the null hypothesis is true. That is, without any adjustment, we will make too many Type I errors, where true null hypotheses are incorrectly rejected. Recall that with $\alpha = 0.05$, one out of twenty tests will result in Type I errors, where a true null hypothesis is rejected. In subsequent chapters we will explore the issue of multiple testing in more detail, but for now we note that the *Bonferroni adjustment* is a simple and straightforward way to account for the problem. To employ this approach, one simply uses α/t instead of α when determining the critical value of the test statistic, where t is the number of tests. For example, if four different grid sizes were used, and we wanted to maintain a Type I error probability of 5%, we would compare each of the four observed test statistics with a chi-square variate having degrees of freedom equal to one less than the number of grid cells, with an area of 0.05/4 = 0.0125 in the upper tail (for a one tail-test of the null hypothesis against the alternative of clustering, which would lead to high values of the statistic). This in turn leads to a higher critical value, and makes it a little more difficult to reject each of the null hypotheses.

5.1.5. Statistical Power

As indicated above, the power of the test will vary depending upon the size and location of the cluster. A cluster of given size and shape will more likely to be "found" by the test if it does not cross quadrat boundaries. Figure 5.2 shows different locations for the same cluster. In (a), the cluster lies entirely within one quadrat, and the test will have greater power – that is, will be more likely to reject the null hypothesis. In (b), it is "diluted" by the fact that it lies partially in four different quadrats.

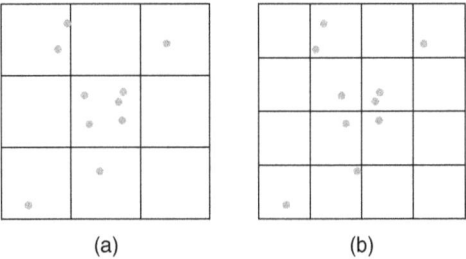

(a)   (b)

Figure 5.2 Identical point distributions with different quadrat configurations

One general way to assess the power of a test is to simulate an *alternative* hypothesis. The simulation is repeated many times, and the proportion of times that the null hypothesis is rejected is then an estimate of the power of the test. To illustrate, 50 points were located in 25 cells as follows: with probability p, a point was located in the cell in the upper left corner cell of a 5×5 grid of cells. With probability $1 - p$, the point was located at random. This will create a cluster in the upper left corner, and the strength of the cluster will depend upon p.

An aside on how to simulate in this way: choose a random number from a uniform distribution on the range $(0,1)$; if it is less than p, the point will be located in the upper left cell; otherwise the point will be chosen at random within the study area. There are several ways to locate a point at random within the study area. Since in this case all cells are the same size, one could simply choose a random number from 1 to 25 (say, by choosing a uniform random number on the range $(0,1)$, multiplying it by 25, taking the integer part of the result, and adding 1). This could also be achieved by assuming a square study area with x- and y-axes on the range $(0,1)$, choosing 50 pairs of x- and y-coordinates on the range $(0,1)$, and then counting the number of x- and y-coordinates that fell within the 25 ranges defined by x- and y-coordinates in the ranges $(0, 0.2)$, $(0.2, 0.4)$, $(0.4, 0.6)$, $(0.6, 0.8)$, and $(0.8, 1)$.

When p is zero, all points are located at random, and the null hypothesis of spatial randomness is true. As p increases away from zero, the likelihood of points clustering in the cell in the upper left corner of the grid increases. The expected number of points in a cell under the null hypothesis is $50/25 = 2$. Under the alternative hypothesis, the expected number of points in the cell in the upper left corner is $50p + 50$ $(1 - p)/25 = 50(p + 1/25 - p/25)$. The simulation was repeated 10,000 times for various values of p. The results are shown in Table 5.1.

As expected, the power of the test increases with the size of the cluster. A plot of power versus p is known as a power curve (Figure 5.3). Power curves illustrate how the ability to reject null hypotheses increases with increasing deviations away from the null hypothesis.

Table 5.1 Changes in statistical power with increasing deviations from null hypothesis

p	No. of points expected in the upper left cell	Power (proportion of null hypotheses rejected)
0	2	0.05 (known as the *nominal* value)
0.05	4.6	0.125
0.1	7.2	0.394
0.125	8.5	0.560
0.15	9.8	0.714
0.175	11.1	0.832
0.2	12.4	0.907

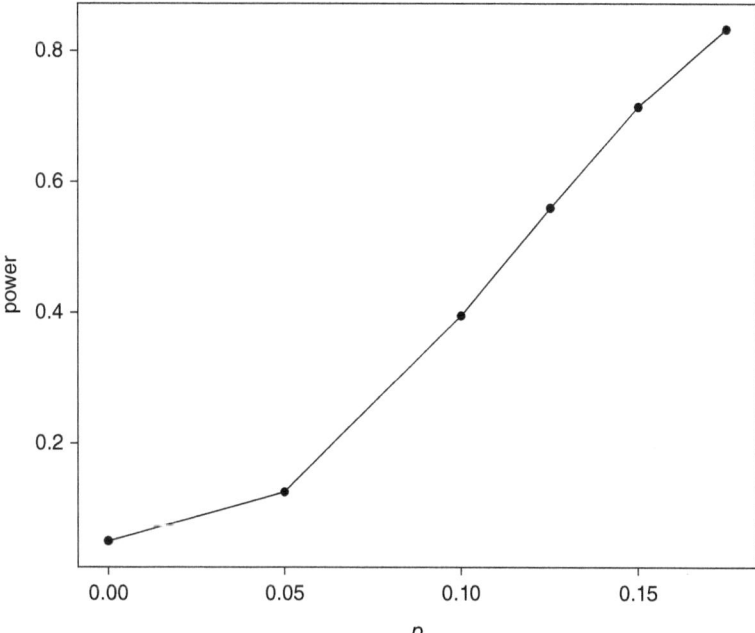

Figure 5.3 Power of quadrat test vs. probability p a point falls in the upper left quadrat

5.1.6 A Further Word on Expectations

Previously we have said little about the expected numbers of cases or events in each region. Here we are assuming that expectations are "given"; in actual applications, there is some underlying model that provides the expected numbers. For instance, a simple model would be to assume equal numbers of events in each region, similar to the original quadrat method discussed at the outset of this section. An alternative model would be to assume that expected numbers were proportional to the population of each region (which in turn might, and would likely, vary from region to region). Or expected numbers might be proportional to the number of older people in the population, if for example the study pertained to the likelihood of having a disease that affected this segment of the population. Another possibility would be to make expectations functions of not only the size and age of the population, but also other characteristics such as income and level of education as well. The question of deciding on expectations is an important one, since those expectations constitute what can be thought of as the "null model." It is the deviations from expectations that will determine the outcome of the hypothesis test. It makes little sense for example to test whether the crime pattern in a city is random – it almost certainly is *not* random. It might make a little more sense to look at crimes per capita, but, realistically, that will likely vary across space as well. Often we would like to know if there is spatial variation, *over* and *above* what we would expect, given known covariates of the dependent variable. Thus if we know that crime depends upon income, age of housing, and other variables, we might wish to predict crime on the basis of those variables (e.g., using regression [see Chapter 8]), and use those predictions as our expectations. Then we can analyze the spatial pattern of the residuals.

5.1.7 Further Limitations

One clear limitation of the quadrat test, like all general/global statistics, is that we simply reject, or fail to reject, the null hypothesis. We don't know where the cluster is, how large it is, what its shape is, etc. Chapter 8, on scan statistics, will address the question of how to identify the location of clusters.

5.2 Moran's *I*

Within geography, Moran's *I* is the most widely used measure of spatial autocorrelation and geographic pattern. The concept underlying Moran's *I* is to assess whether high values on a map are located near other high values (and whether low values are located near other low values). This is seen through the equation for the statistic:

$$I = \frac{n\sum_{i=1}^{n}\sum_{j=1}^{n}w_{ij}\left(x_i - \bar{x}\right)\left(x_j - \bar{x}\right)}{\left(\sum_{i=1}^{n}\sum_{j=1}^{n}w_{ij}\right)\sum_{i=1}^{n}\left(x_i - \bar{x}\right)^2} \tag{5.5}$$

where *n* is the number of regions, x_i is the value of the variable of interest, and w_{ij} is a weight describing the relationship between region *i* and region *j*.

Although this is perhaps a daunting equation, it can be understood fairly easily, as follows. First simply note that this is a measure of spatial *auto*correlation – we are examining the correlation of a variable at a location with itself, at nearby locations. It is worth noting that the equation above is similar in structure to the familiar correlation coefficient (Pearson's *r*):

$$r = \frac{\sum_{i=1}^{n}(x_i - \bar{x})(y_i - \bar{y})}{\sqrt{(x_i - \bar{x})^2}\sqrt{(y_i - \bar{y})^2}}$$ (5.6)

The difference is that we are now correlating a variable with itself, in nearby regions.

The conceptually important part of the equation for Moran's *I* is the numerator. For all pairs of regions, the product of the deviations from the mean $(x_i - \bar{x})(x_j - \bar{x})$ is calculated. Note that this will be positive when values at regions *i* and *j* are both above (or both below) the mean. In this case, the pair of regions is contributing to *positive* spatial autocorrelation. The term will be negative when one of the two regions is above the mean, and the other is below the mean; in these cases the pair of regions is contributing towards *negative* spatial autocorrelation. Although in practice measures of spatial autocorrelation are usually positive, Griffith (2019) notes that particular examples often contain mixtures of positive and negative spatial autocorrelation, and he argues that more attention should be paid to the negative spatial autocorrelation that is often hidden by the global measures that are positive.

The denominator serves two functions. First, the sum of the weights is used to make the quantity a weighted average – quantities in the numerator are weighted, and then a weighted average is found (as it always is) by dividing by the sum of the weights. The denominator also contains the quantity $\sum_{i=1}^{n}(x_i - \bar{x})^2 / n$ (since the *n* here is in the denominator of a denominator, it appears in the numerator of the equation). This is of course the (biased version of the) sample variance, and it serves to standardize the covariance in the numerator, placing the statistic on a scale ranging from −1 to +1. Values of *I* near zero occur when the variable is distributed randomly across the map.

The weights serve to attach an importance to the relative locations of *i* and *j*. Generally, the weight w_{ij} is high when *i* and *j* are near to one another, and low when the two regions are far apart. By convention the weights w_{ii} are equal to zero; the relationship of a region with itself is ignored, and does not contribute to the calculation. A common choice for the definition of weights is one of *binary adjacency*. Here the weight is equal to one if regions *i* and *j* are adjacent, and zero if they are not adjacent. Adjacency itself needs to be defined. Perhaps the most common definition is to define two regions as adjacent if they share a common border of non-zero length. This is known as rook's adjacency. Adjacency can also be defined for any regions that touch at a point (i.e., it is not necessary to have a non-zero length boundary); this is known as queen's adjacency. Throughout the remainder of the book, rook's adjacency is used. It is more commonly employed than queen's adjacency, perhaps because it is a stricter version of "adjacent," since a border must be shared. In any event, the researcher should use a definition that is consistent with the best educated guess regarding the interregional relationships that characterize individual application. This will be discussed in more detail later.

Another way to simplify the equation is via the use of z-scores:

$$I = \frac{n\sum_{i=1}^{n}\sum_{i=1}^{n}w_{ij}z_jz_j}{\left(\sum_{i=1}^{n}\sum_{i=1}^{n}w_{ij}\right)} \qquad (5.7)$$

A simple example is shown for the five-region system in Figure 5.4. The respective values for the five regions are (6, 7, 8, 2, and 2). Using binary adjacency, the weights are equal to one for the following twelve pairs: (1,2), (1,3), (2,1), (2,4), (2,5), (3,1), (3,4), (4,2), (4,3), (4,5), (5,2), and (5,4). Note that pairs of regions are counted twice (e.g., the connection between regions 1 and 2 is captured by the pairs (1,2) and (2,1)). Weights for all other pairs are equal to zero.

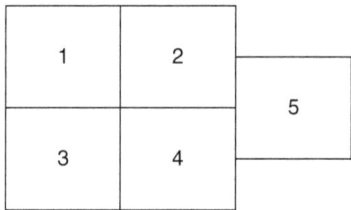

Figure 5.4   Spatial configuration of a hypothetical five-region system

The calculation proceeds as follows. The mean is equal to 5. In the numerator we add the products of deviations from the mean for those pairs of regions that are adjacent to one another (the other pairs are ignored, since their weight is equal to zero). These products are each multiplied by the weight, which is equal to one. Thus the numerator is equal to $n = 5$, multiplied by

$$
\begin{aligned}
&(6{-}5)(7{-}5) + (6{-}5)(8{-}5) + (7{-}5)(6{-}5) + (7{-}5)(2{-}5) + (7{-}5)(2{-}5) +(8{-}5)(6{-}5) + (8{-}5) \\
&(2{-}5) + (2{-}5)(7{-}5) + (2{-}5)(8{-}5) + (2{-}5)(2{-}5) + (2{-}5)(7{-}5) + (2{-}5)(2{-}5) \\
&= 2 + 3 + 2 - 6 - 6 + 3 - 9 - 6 - 9 + 9 - 6 + 9 = -14 \qquad (5.8)
\end{aligned}
$$

Note that a shortcut to the calculation of the numerator is to calculate the product once for each of the six pairs of regions (where we *don't* double count), and then multiply the result by 2.

The denominator is equal to the sum of the weights (12) multiplied by the sum of the squared deviations from the mean:

$$(6{-}5)^2 + (7{-}5)^2 + (8{-}5)^2 + (2{-}5)^2 + (2{-}5)^2 = 1 + 4 + 9 + 9 + 9 = 32 \qquad (5.9)$$

Moran's I is equal to $\{5(-14)\}/\{12 \ (32)\} = -70/384 = -0.182$.

Although it is common to use binary adjacency as a definition for the weights, this is by no means the only way to define them. They could be defined for example in terms of reciprocal distance ($w_{ij} = 1/d_{ij}$ where d_{ij} is the distance between the centroids of regions i and j). But more important is a consideration of the process and application that is being studied. The choices described above imply a specific spatial scale for the dependency. It may be that region i influences, and is influenced by, regions that are far away from it. It may be that instead of reciprocal distance, influence falls off more sharply, perhaps for example with the square of distance. The definition of weights also does not necessarily have to be a symmetric one. The point here is that the researcher should think carefully about the definition of the weights – different definitions will give rise to different values of Moran's I. In some cases, an exploratory approach may be appropriate, where different weights are tried to assess how sensitive the result is to differing definitions. In an hypothesis testing framework, if a range of definitions is tried, we will want to account for the multiple testing. In subsequent sections, we will address these issues in more detail.

5.2.1 Hypothesis Testing

How can we test the null hypothesis of spatial randomness? How high (or low, in the case of negative spatial autocorrelation) does Moran's I have to be before we reject the null hypothesis? To construct a test, we need to know more about the null distribution of the statistic – that is, the distribution of Moran's I when the null hypothesis is true.

The expectation of Moran's I is not exactly equal to zero; it is slightly negative, and equal to

$$E[I] = \frac{-1}{n-1} \tag{5.10}$$

where n is the number of regions. To construct a z-score and carry out a z-test, we also need to know how variable I is when the null hypothesis is true. One approximation for the variance is $V[I] = 1/(2n)$ for the case of a set of square grid cells, using binary and rook's adjacency. In this case each region is connected to four other regions. For a general regional system with binary adjacency on a typical map (e.g., a map of the US states), an approximation for the variance is $1/(3n)$.

The expressions for the expectation and variance allow construction of a z-score and hypothesis testing. If the z-score is higher than its critical value, the null hypothesis of spatial randomness is rejected in favor of the conclusion of positive spatial autocorrelation.

There are two more exact expressions for the variance that are available. They differ in terms of the null process that is used to construct the null distribution. These can be understood by first imagining two different types of simulation of spatial randomness. One is based upon *random permutations*. With a set of values $\{x_1, ..., x_n\}$ on a map, imagine taking all of the values off of the map, and then randomly tossing them back onto the map. Then compute Moran's I, and after that, repeat this many times, creating a distribution of Moran's I. We can use, for example, the 95th percentile of this distribution as the critical value for I. (In addition, the simulated distribution may or may not be normal, and so this process would address another weakness of using the z-test described above – namely, that the distribution of Moran's I is not precisely a normal distribution.)

Another way to simulate the null hypothesis is to assume that the value in each region comes from an identical normal distribution with the same mean and variance. Using the sample mean and sample variance, one could sample from a normal distribution with those parameters, and create simulated values for each of the n regions. This is the assumption of *normality*. Moran's I would then be calculated, and the process repeated many (often 1,000 or 10,000) times, and again this distribution would serve as a null distribution from which critical values can be established. The equations for the variance of I under the null hypothesis are as follows.

Assumption of normality:

$$V[I] = \frac{1}{(n-1)(n+1)S_0^2}\left(n^2 S_1 - n S_2 + 3 S_0^2\right) - \left(\frac{-1}{n-1}\right)^2 \tag{5.11}$$

Assumption of equally likely random permutations:

$$V[I] = \frac{n\{(n^2 - 3n + 3)S_1 - n S_2 + 3 S_0^2\} - b_2\{(n^2 - n)S_1 - 2n S_2 + 6 S_0^2\}}{(n-1)(n-2)(n-3)S_0^2} - \left(\frac{-1}{n-1}\right)^2 \tag{5.12}$$

where

$$b_2 = \frac{m_4}{m_2^2} \tag{5.13}$$

$$m_4 = \frac{\sum_{i=1}^{n} z_i^4}{n} \tag{5.14}$$

$$m_2 = \frac{\sum_{i=1}^{n} z_i^2}{n} \tag{5.15}$$

$$S_0 = \sum_{i=1}^{n}\sum_{j \neq i}^{n} w_{ij} \tag{5.16}$$

$$S_1 = \frac{1}{2} \sum_{i=1/j\neq i}^{n}\sum^{n} \left(w_{ij} + w_{ji}\right)^2 \qquad\qquad (5.17)$$

$$S_2 = \sum_{i=1}^{n} \left(\sum_{j=1}^{n} w_{ij} + \sum_{j=1}^{n} w_{ji} \right)^2 \qquad\qquad (5.18)$$

In the various routines and packages that are used to assess Moran's I, the random permutations assumption is probably more widely assumed than the normality assumption.

Now imagine a situation where we are interested in the spatial pattern of a disease across space, and we have data for the zip or postal codes for a metropolitan area. For each spatial unit we have a mortality rate, and we would like to test the null hypothesis that these rates have no spatial pattern. Let's think about the two simulations just described. In the first case, we take all of the rates off of the map and put them back on, and this is done repeatedly. When this is done, there is an implicit assumption that each of those random permutations is equally likely. This, however, may not be the case in practice. Some of the zip/postal codes have higher populations than the populations found in others. The rates in the low-population areas are more variable than the rates in the high-population areas. Every rearrangement of the rates on the map is therefore *not* equally likely. For example, a rearrangement of the regional values that has very high and/or very low values in say the central city where population densities are high will be less likely than a permutation where the extreme, high, and/or low rates are found on the periphery, where populations are low. Thus the use of random permutations to establish the null distribution of Moran's I is often not a good idea.

What about the normality assumption – can that be used to create the null distribution? For the same reason as just described, this assumption is often problematic as well. Stimulating rates in each region using an identical normal distribution with the same mean and variance would not mimic well the actual situation. Disease rates in the dense, highly populated regions would have a *lower* variance – rates there are more stable because of their higher denominators. Similarly, rates in the lightly populated regions would be expected to be more variable (i.e., have higher variances). Thus the assumption of normality also is questionable when testing the null hypothesis.

So, what can be done? We know the average mortality rate, and we know the number of people in each of the regions. One possibility is to simulate cases in each region. If, for example, we have a total population of P, and C is the total number of cases, then the overall disease rate is C/P. We could choose random numbers and assign each person in each region the disease with probability C/P. This would have the desired effect of creating relatively more (less) variable mortality rates in the low- (high-) population regions. Moran's I could be calculated, and this could be repeated many times to establish a more realistic distribution of the statistic when the null hypothesis of spatial randomness holds.

5.3 Geary's C

Geary's C is an alternative measure of spatial autocorrelation (Geary, 1954) and it has a form that is a bit similar to Moran's I; there is a numerator and a denominator, and there are weights:

$$C = \frac{(n-1)\sum_{i=1}^{n}\sum_{j=1}^{n}w_{ij}\left(x_i - x_j\right)^2}{2(\sum_{i=1}^{n}\sum_{j=1}^{n}w_{ij})\sum_{i=1}^{n}\left(x_i - \bar{x}\right)^2} \tag{5.19}$$

The conceptual core of the statistic is the term $\left(x_i - x_j\right)^2$ in the numerator. The focus is on the difference between two nearby values. If this difference is small, the numerator and the statistic will be small. The lower limit of C is equal to zero, and small values of C near zero therefore imply positive spatial autocorrelation. When x_i and x_j are very different the numerator and C will be large; C is scaled to have an upper limit of two, and values of C near this characterize examples of negative spatial autocorrelation. Random patterns will have a value of C near its expected value of one.

Hypothesis testing can be carried out with knowledge of the variance of C, under the null hypothesis. There are again two ways to generate the null distribution – through the assumption of normality, or through the assumption of equally likely random permutations. The variances of the statistic under the null hypothesis, for these assumptions, are as follows:

Assumption of normality:

$$V[C] = \frac{(2S_1 + S_2)(n-1) - 4S_0^2}{2(n+1)S_0^2} \tag{5.20}$$

Assumption of random permutations:

$$V[C] = \frac{(n-1)S_1\left\{n^2 - 3n + 3 - (n-1)b_2\right\}}{n(n-2)(n-3)S_0^2} - \frac{(n-1)S_2\left\{n^2 + 3n - 6 - \left(n^2 - n + 2\right)b_2\right\}}{4n(n-2)(n-3)S_0^2} + \frac{n^2 - 3 - (n-1)^2 b_2}{n(n-2)(n-3)} \tag{5.21}$$

where b_2, S_0, S_1, and S_2 are defined in Equations 5.13 and 5.16–5.18.

5.3.1 Comparison of C and I

One interesting question that arises is, "which measure is better?" What does this question even mean? One interpretation can be given in terms of statistical power – the measure that is "better" is the one that rejects the null hypothesis of spatial randomness more often when the null hypothesis is false. In this regard, it turns out that Moran's I is generally more powerful against a broad range of alternatives. And in fact Moran's I is certainly used more often than Geary's C. But are there circumstances – certain types of situations – where it might be preferable to use Geary's C?

Geary's C appears to have slightly greater power, in comparison with Moran's I, when it comes to rejecting the null hypothesis that a map of p-values is spatially random (see, e.g., Rogerson and Yamada, 2009).

For example, a map of cancer rates could be assessed for patterns. But we know that denominators are also important – small regions might have high cancer rates, but because of the small population, the associated p-value (say from a hypothesis test that the proportion of people with cancer was no different than, for example, the statewide or nationwide rate) could be insignificant. Thus we have an interest in looking at maps of p-values in addition to maps of rates. Is there clustering on maps of p-values? Are low p-values located near to other low p-values? If the null hypothesis that there is no raised rate in a specific region is true, then p has a uniform distribution between 0 and 1 – every possible value of p is equally likely. Geary's C appears to be slightly better able (in comparison to Moran's I) to detect clusters of low p-values.

For both I and C, weights defined for pairs of places can alternatively be defined as a function of distance between places. For example, $w_{ij} = d_{ij}^{b}$, where b is a parameter reflecting how rapidly the weight declines with increasing distance (another example would be to define $w_{ij} = e^{-d_{ij}^{b}}$). One could explore the importance of the role of distance in spatial autocorrelation by trying several different specifications for the weights, using different b-values. This would lead to different degrees of spatial autocorrelation, and different p-values associated with hypothesis tests designed to detect departures from randomness. Significant departures from randomness might be found for some spatial scales but not others. In cancer mapping, spatial dependence is often *not* found when the units of analysis are counties. Counties are large and have a lot of internal heterogeneity. The rate of cancer in one county is not necessarily similar to the rate in another county. If, however, the units of analysis are census blocks, it is more likely that the rate in one block will be similar to the rate in an adjacent block.

5.4 Tango's Statistic

Tango (1995) developed a global statistic that is a measure of spatial pattern:

$$C_G = \sum_{i=1}^{m} \sum_{j=1}^{m} w_{ij} (r_i - p_i)(r_j - p_j) \tag{5.22}$$

where r_i is the proportion of all cases falling in region i, and p_i is the proportion of all cases expected to fall in region i. It looks at pairs of regions and compares what is observed with what is expected. The equation resembles Moran's I to some degree – the sum is over all pairs of regions, and these pairs are weighted. One difference is that there is no standardizing denominator.

Tango's statistic may be disaggregated:

$$C_G = \sum_{i=1}^{m} w_{ii} \left(r_i - p_i \right)^2 + \sum_{i=1}^{m} \sum_{j=1 j \neq i}^{m} w_{ij} \left(r_i - p_i \right) \left(r_j - p_j \right) \tag{5.23}$$

In this form it is clear that Tango's statistic is a combination of (a) the first term on the right-hand side above, which is a chi-square-like statistic that captures what is going on within regions, and (b) a Moran-like statistic that captures the product of excess above expectations for pairs of regions. By convention, Moran's I does not capture what is occurring *within* regions (w_{ii} is equal to zero). Similarly the chi-square goodness-of-fit statistic does not capture what is occurring between regions that are nearby to one another.

In matrix form

$$C_G = (\mathbf{r} - \mathbf{p})' \, \mathbf{W}(\mathbf{r} - \mathbf{p}) \tag{5.24}$$

where $\mathbf{r}$ and $\mathbf{p}$ are, respectively, $m \times 1$ column vectors of the proportion of observed and expected cases in each region. $\mathbf{W}$ is the $m \times m$ matrix containing the weights, $\{w_{ij}\}$. (The bolded notation is used for matrices and vectors; for more details on matrix algebra, see Appendix A).

Tango's statistic will be high when either the first term on the right-hand side is high, or the second term on the right-hand side is high, or both. We are able to reject the null hypothesis of spatial randomness when either (a) observations in regions deviate substantially from expectations, and/or (b) regions that have observations that are higher (lower) than expected are near regions that also have observations that are higher (lower) than expectations.

5.5 A Spatial Version of the Chi-Square Statistic

The spatial chi-square statistic (Rogerson, 1999) is written as

$$R = \sum_{i=1}^{m} \sum_{j=1}^{m} w_{ij} \left(\frac{\left(r_i - p_i \right)}{\sqrt{p_i}} \right) \left(\frac{\left(r_j - p_j \right)}{\sqrt{p_j}} \right) \tag{5.25}$$

or, equivalently,

$$R = \sum_{i=1}^{m} w_{ii} \frac{\left(r_i - p_i \right)^2}{p_i} + \sum_{i=1}^{m} \sum_{j=1 j \neq i}^{m} w_{ij} \left(\frac{\left(r_i - p_i \right)}{\sqrt{p_i}} \right) \left(\frac{\left(r_j - p_j \right)}{\sqrt{p_j}} \right) \tag{5.26}$$

The first term on the right-hand side is a chi-square statistic, where the contributions from each region i are weighted by w_{ii}. In matrix form,

$$R = (\mathbf{r} - \mathbf{p})' \, \tilde{\mathbf{W}} (\mathbf{r} - \mathbf{p}) \tag{5.27}$$

where $\tilde{\mathbf{W}}$ is a modified weights matrix containing elements $w_{ij} / \left(\sqrt{p_i} \sqrt{p_j} \right)$. The spatial chi-square statistic therefore is simply a special case of Tango's statistic, where the weights are modified by dividing by the square root of the product of expected proportions in regions i and j. Like Tango's statistic, the spatial chi-square statistic will be high and possibly significant when the first and/or second terms on the right-hand side of Equation 5.26 is/are high.

In comparing the Tango and spatial chi-square statistics, the relative power of these two tests is such that the spatial chi-square test is relatively more powerful when one has deviations from expectations in small regions. On the other hand, Tango's statistic has relatively higher power (in comparison to the spatial chi-square statistic) when there are substantial deviations from expectations in large regions. For example, when considering a county-based map of New York State, Tango's statistic will be better at rejecting false null hypotheses when the contribution to deviations from expectations comes from large regions such as New York City, Rochester, and Buffalo. When there are substantial deviations from expectations in small counties (e.g., in the Adirondack Mountains of New York State), the spatial chi-square test will generally be better.

Figure 5.5 shows a regional system with two hypothetical sets of values for a variable of interest. In part (b), the deviations from expectations are the same as those in part (a), but they are arranged across regions differently. If one were to carry out an aspatial chi-square goodness-of-fit test comparing observed and expected values, the maps in both parts of the figure would yield the same chi-square statistic. However, when one looks at the maps, it is easy to see that the high deviations are near one another in part (a). The common aspatial chi-square goodness-of-fit test does not take into account the spatial configuration and spatial relationships between regional values – it only uses information about what is observed and expected in individual regions.

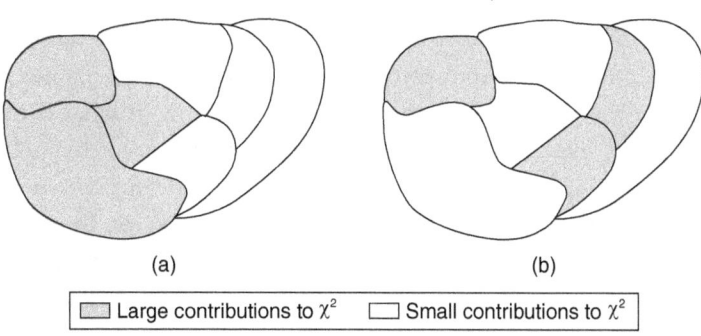

(a)   (b)

☐ Large contributions to χ^2 ☐ Small contributions to χ^2

Figure 5.5a,b Two patterns of deviations (observed – expected) yielding identical chi-square values

Source: Rogerson (1999). Reprinted with permission of John Wiley and Sons.

Figure 5.6 depicts another example. Suppose the values shown are residuals from a regression analysis, capturing the difference between observations and the values predicted by the regression (regression analysis is discussed in detail in many introductory texts on statistics). Notice that the values add to zero; this is a property of residuals. There are errors in each region. Some of these errors are positive, reflecting the case where observed values of the dependent variable are greater than that predicted by the regression, and some errors are negative, reflecting overprediction by the regression. Comparing part (a) of the figure with part (b), we see that in the latter case, the regression does not work as well – the errors are relatively higher. Now suppose we are interested in the spatial pattern of residuals, and we use Moran's I to measure that pattern. The pattern of the residuals is the same, and the computed value of Moran's I will be the same in both (a) and (b). Here we see that Moran's I is a statistic that focuses upon spatial pattern, but does not focus upon the magnitude of errors within regions – the maps do have identical patterns, but one map (namely, map b) has much larger deviations from what is expected or predicted by the regression.

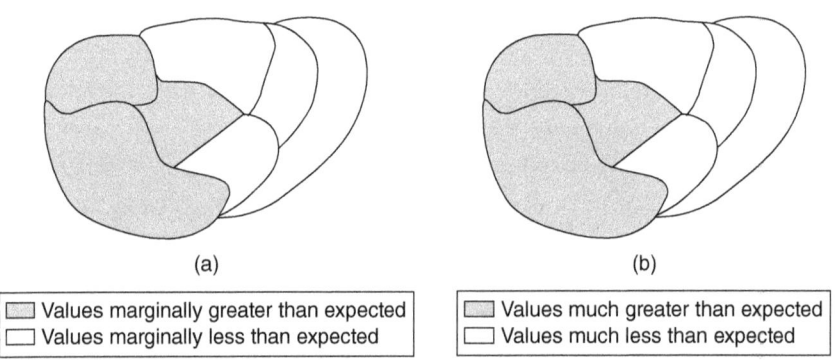

(a)	(b)
▨ Values marginally greater than expected	▨ Values much greater than expected
☐ Values marginally less than expected	☐ Values much less than expected

Figure 5.6a,b   Two patterns of deviation (observed – expected) yielding identical values of Moran's I

Source: Rogerson (1999). Reprinted with permission of John Wiley and Sons.

Tango's statistic and the spatial chi-square statistic are aimed at combining these two situations. Each statistic is the sum of two parts, where one part focuses upon what is occurring within regions (by comparing what is observed with what is expected). The other part is a Moran-like statistic that focuses upon what is occurring between pairs of regions. The statistics approach and reach statistical significance when they are sufficiently high, and this can occur because either (or both) of these two measured components are high. When the global statistic is high, it is interesting to decompose it into its chi-square-like and Moran-like components (Equations 5.23 and 5.26), to see where the contributions are coming from.

5.6 Significance of Tango's Statistic and the Spatial Chi-Square Statistic

When the null hypothesis of spatial randomness holds, what are the distributions of these two statistics? The answer to this question is important because it will help us to determine critical values for the statistics. When the value of the statistic is high enough, we wish to reject the null hypothesis. We therefore need to know how unusual our statistic would be, if indeed the null hypothesis were true.

For Tango's statistic, the expected value of C_G is equal to

$$E[C_G] = \frac{1}{N} \mathrm{Tr}\left(\mathbf{W}\mathbf{V}_p\right) \tag{5.28}$$

where N is the number of cases, "Tr" refers to the trace of a matrix (and, in turn, the trace is equal to the sum of the diagonal elements), and where

$$V_p = \Delta\mathbf{p} - \mathbf{p}\mathbf{p}' \tag{5.29}$$

and $\Delta\mathbf{p}$ is equal to a $m \times m$ diagonal matrix, with elements $p_1, p_2, ..., p_m$ along the diagonal, and zeros elsewhere. When the null hypothesis is true, the variance of Tango's statistic is

$$V[C_G] = \frac{2}{N^2} \mathrm{Tr}\left(\mathbf{W}\mathbf{V}_p\right)^2 \tag{5.30}$$

To test the hypothesis of spatial randomness, we can assume that C_G is normally distributed, with mean and variance given by Equations 5.28 and 5.30. Then the z-statistic

$$z = \frac{C_G - E[C_G]}{\sqrt{V[C_G]}} \tag{5.31}$$

has, approximately, a standard normal distribution with mean 0 and variance 1. For 95% of the time, when the null hypothesis is true, this statistic will fall between −1.96 and +1.96. The use of this normal distribution is only an approximation. A better approximation for the null distribution of C_G is a chi-square distribution. Details are provided in Tango (1995) and Rogerson (1999).

The spatial chi-square statistic has the same distribution; the only difference is that the modified weights $\tilde{\mathbf{W}}$ described above are used in place of $\mathbf{W}$.

A comparison of these two statistics is given by Rogerson (1999). Summarizing that comparison, the circles on the map in Figure 5.7 are proportional to the hypothetical populations of 20 places. Simulations were carried out, where N cases were distributed across the population. Raised risk was assumed at the center of a chosen population center, and an exponential decline in risk with distance was used. Thus most places were assigned cases randomly, using the overall rate for the region. One place was more likely to receive cases, and the surrounding places, if close enough, were also slightly more likely to receive additional cases. Table 5.2 shows the statistical power of each test, for the 20 different scenarios (corresponding to raised risk in each of the 20 places). The simulation for each scenario was repeated 1,000 times, and the entries in the table show the proportion of times that the false null hypothesis of spatial randomness was rejected. These entries are therefore equal to the statistical power of the test.

Table 5.2   Comparison of power for global clustering statistics

Location	R (normal)	C_G	χ^2
1	.653	.334	.517
2	.890	.996	.887
3	.480	.315	.516
4	.383	.077	.312
5	.825	.968	.772
6	.733	.383	.563
7	.668	.564	.634
8	.848	.984	.797
9	.480	.322	.515
10	.749	.890	.762
11	.744	.421	.608
12	.733	.740	.613
13	.482	.272	.402
14	.739	.746	.641
15	.195	.059	.189
16	.853	.925	.827
17	.140	.049	.145
18	.421	.090	.387
19	.283	.063	.253
20	.431	.360	.388

Source: Rogerson (1999). Reprinted with permission of John Wiley and Sons.

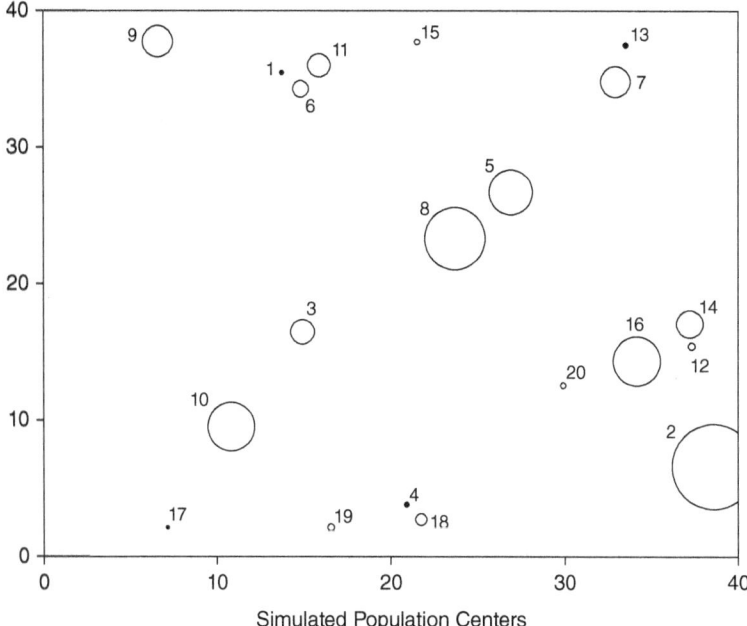

Figure 5.7 Simulated population centers

Source: Rogerson (1999). Reprinted with permission of John Wiley and Sons.

The general conclusion is that the spatial chi-square test is better at finding raised risk in small locations, while Tango's statistic has relatively higher statistical power when there is raised risk in locations with higher populations. When there are no spatial effects (e.g., when risk is raised in Location 3, which is isolated spatially) – that is, there is no spillover of risk from one location to surrounding locations – then the aspatial chi-square test works best. When there are no spatial effects, there is no reason or need to use spatial methods to look for such effects; those spatial methods actually will yield lower powers than aspatial methods. When there is a small location with spillover of risk to a surrounding location or locations (as is the case when increased risk is centered on Locations 1, 6, and 11, for example), then the spatial chi-square test has highest power. Tango's statistic has only 33% power for Location 1, while the spatial chi-square statistic has power equal to about 65% in this case. In Location 8 the power of the Tango statistic is 0.984; here the power of the spatial chi-square statistic is about 0.80 to 0.85. Locations 2 and 10 also have large populations, and again the Tango statistic outperforms the spatial chi-square statistic.

Note too that power is generally lower in small locations than it is in large locations – it is intuitively more difficult to find clusters in smaller locations, where there are fewer cases.

When using these global statistics, the choice of weights can make a difference. In choosing the weights, the analyst is essentially estimating the spatial scale of the process (in this case, the spatial extent of the increased risk). If the analyst chooses weights associated with either (a) a small cluster when risk is actually more extensive spatially, or (b) a large cluster when in fact the cluster is not that large, statistical power will not be as high as it would be if the weights matched the actual cluster size. This is known as the matched-filter theorem – the best chance of finding a cluster occurs when using a statistic with weights that match the size of the cluster. This is shown in Table 5.3, where τ ("tau") is a parameter associated with the size of the cluster. The simulations of clusters were carried out using $\tau = 5$. But the analyst of course will not know how large the actual cluster is. A choice of weights has to be made, and here this is done through the choice of τ, which is directly proportional to the size of the cluster that is expected.

Table 5.3   Power of R for different values of τ at selected locations

| | Location | | | |
τ	3	6	7	12
0.1	.516	.563	.636	.613
1	.513	.647	.645	.639
2	.492	.695	.649	.669
3	.472	.704	.641	.693
4	.453	.706	.630	.701
5	.434	.704	.619	.702
6	.412	.696	.611	.695
7	.393	.691	.603	.684
8	.374	.682	.593	.673
9	.357	.673	.588	.664
10	.339	.667	.580	.649
11	.324	.662	.572	.639
12	.316	.656	.565	.629
13	.302	.648	.557	.619
14	.288	.641	.547	.610
15	.276	.633	.538	.600
16	.264	.630	.531	.590
17	.256	.624	.527	.576
18	.247	.617	.516	.565
19	.237	.610	.506	.554
20	.227	.606	.498	.548

In Location 3, using the chi-square statistic, you will find a cluster (i.e., reject the null hypothesis) 51% of the time. If you look for a larger cluster (by choosing weights via choices of $\tau > 0$), power is reduced below 51%. Now let's examine Location 6. Location 6 is a relatively small place, and it is located near Locations 1 and 11. When cases are assigned in the simulation, Locations 1 and 11 will also receive excess cases. Power increases with the choice of τ as τ increases from 0 to about 4; the power increases from 0.563 to a maximum of 0.706. If you happened to guess that the cluster was of a size equivalent to $\tau = 4$, the likelihood of finding the cluster would be quite high (0.706). If spatial effects were ignored, a cluster would be found at Location 6 only about 56% of the time. Similarly, if weights were chosen assuming a cluster of size larger than simulated (as τ increases beyond 5), power would drop with increasing τ. This is a good illustration of the matched-filter theorem – statistical power is maximized when the weights are chosen to be associated with a cluster size that is similar to the actual cluster size. When the cluster is centered on Location 12, again we find that power is maximized at about $\tau = 5$.

Are we better off guessing a cluster size that is too small, or a cluster size that is too large? If the profile of power vs. τ is as in Table 5.3, then it would be OK if the estimate of cluster size was too large – power does not drop substantially as τ increases beyond the actual cluster size.

The reader should also understand that the power values shown in Tables 5.2 and 5.3 are a function of the number of cases and the amount of raised risk. The power of each statistic would be larger if, for example, the magnitude of raised incidence was greater within the simulated clusters, or if the total number of cases was greater. Some simulations of statistical power that compare statistics choose parameters in such a way that power approaches 95% to 100%. But such choices limit the usefulness of the simulation for comparing statistics, since there is not much difference between two statistics displaying powers of, say, 96% and 98%. In practice, choosing simulation powers to find a midrange for power (not near 0–10% or 90–100%) will be more informative for comparisons.

Numerical example of Tango's statistic

Suppose we have a four-region system with the following weight matrix:

$$\mathbf{W} = \begin{bmatrix} 1 & 0.5 & 0.5 & 0 \\ 0.5 & 1 & 0.3 & 0.3 \\ 0.5 & 0.3 & 1 & 0.5 \\ 0 & 0.3 & 0.5 & 1 \end{bmatrix}$$

Tango's statistic has two components. One is the sum for each region of observed minus expected values, squared. And for pairs of regions, we want the product of the deviations of observations from expectations.

Suppose that the observed proportion of cases in each of the four regions is $\mathbf{r}' = \{0.4\ 0.3\ 0.1\ 0.2\}$. Suppose also that the expectations are $\mathbf{p}' = \{0.3\ 0.2\ 0.2\ 0.3\}$. These expectations come from somewhere; here they are merely "given". In practice, these might come from, in a simple case, the proportion of population that lives in each region. They might be the proportion of the elderly population that lives in each region, if we wish to restrict our attention to that subset of the population. They might be predictions made from a regression model that for example estimates the number of cases based upon variables such as age structure, income and education levels, percent minority, etc.

We can compute the statistic $C_G = (\mathbf{r}-\mathbf{p})'\mathbf{W}(\mathbf{r}-\mathbf{p})$, using $(\mathbf{r}-\mathbf{p}) = \{0.1\ 0.1\ -0.1\ -0.1\}$. Or, equivalently, we could use the algebraic formula given in Equation 5.22.

Next, we can compare this observed value with what is expected when the cases are distributed randomly, according to expectations.

We have

$$\Delta\mathbf{p} = \begin{bmatrix} 0.3 & 0 & 0 & 0 \\ 0 & 0.2 & 0 & 0 \\ 0 & 0 & 0.2 & 0 \\ 0 & 0 & 0 & 0.3 \end{bmatrix}$$

Subtracting $\mathbf{pp}'$ from this yields

$$V_p = \begin{bmatrix} 0.21 & -0.06 & -0.06 & -0.09 \\ -0.06 & 0.16 & -0.04 & -0.06 \\ -0.06 & -0.04 & 0.16 & -0.06 \\ -0.09 & -0.06 & -0.06 & 0.21 \end{bmatrix}$$

Premultiplying this by the weight matrix, $\mathbf{W}$, yields a result with the elements 0.15, 0.10, 0.088, and 0.162 along the diagonal. The sum of these is the trace, and it is equal to 0.5. With $N = 100$ cases, we expect Tango's statistic to be, on average, $E[C_G] = (1/100)(0.5) = 0.005$. We also want the variance; to find this we need to first find the 4 x 4 matrix $(\mathbf{WV}_p)(\mathbf{WV}_p)$. The trace of this matrix is equal to 0.1156.

We now compare our observed statistic with what was expected, calculating the z-score:

$$z = \frac{C_G - E[C_G]}{\sqrt{V[C_G]}} = \frac{0.038 - \left(\frac{1}{N}\right)0.5}{\sqrt{(2/N^2)(0.1156)}} = \frac{0.038N - 0.5}{\sqrt{2(0.1156)}} \tag{5.32}$$

If for example $N = 20$, $z = \{0.038(20) - 0.5\}/\sqrt{2(0.1156)} = 0.54$. The observed value of Tango's statistic (0.038) was a little higher than expected (0.5/20 = 0.025). However, our z-score is not far enough from zero to reject the null hypothesis. If $N = 100$ then $z = \{0.038(100) - 0.5\}/\sqrt{2(0.1156)} = 6.86$, and this is highly significant – we wouldn't expect the results we observed by chance alone. If N is around 40, then z is close to 2, and on the margin of statistical significance.

5.6.1 COVID-19 Cases in a Region of New York State

Table 5.4 shows the observed number of COVID-19 cases in a subset of counties of the state of New York State for March 28, 2020. The expected number of cases is derived assuming a uniform risk throughout the population. Thus the expected number of cases in a county is equal to the total number of cases (42), multiplied by the county's share of the total population in the seven-county region.

Table 5.4 COVID-19 cases in central New York State

	County	Cases (as of 3/28/20)	Population	Expected Cases	z
1	Otsego	3	59749	6.95	−1.5
2	Schoharie	2	31097	3.62	−0.85
3	Delaware	7	44527	5.18	0.80
4	Greene	5	47491	5.52	−0.22
5	Columbia	13	59916	6.97	2.28
6	Chenango	3	47536	5.53	−1.08
7	Madison	9	70795	8.23	0.268
	Total	42	361,111	42	

The z-score in the last column is equal to cases minus expected cases, divided by the square root of the expected number of cases.

The chi-square goodness-of-fit test comparing observed and expected case numbers yields a statistic of 10.105; this is lower than the critical value of 12.59 found using $\alpha = 0.05$ and $7 - 1 = 6$ degrees of freedom. The p-value of 0.12 implies that about 12% of the time, we would expect a higher value of the statistic, when the null hypothesis is true. The null hypothesis of no raised risk across regions cannot be rejected.

To calculate Moran's I for a map of the z-scores, we first need to know which counties are connected to which. Using rook's case binary adjacency, the following pairs are adjacent: {(1,2) (1,3) (1,6) (1,7) (2,3) (2,4) (3,4) (4,5) (6,7)}. The total number of entries in the **W** matrix is double this number, or 18. Thus the sum of the weights is $9 \times 2 = 18$, since, for example, there is a "(1,2)" pair and a "(2,1)" pair.

The mean of the z's is –0.0437. The sum of the squared deviations from the mean is 10.085. The conceptual part of the numerator – the sum of the products of adjacent deviations from the mean – is equal to –0.8469. Moran's I is therefore equal to $(7)(–0.8469)/\{18 \times 10.085\} = –0.0327$. Since the expected value of Moran's I is equal to $–1/(7 - 1) = –0.1666$, this indicates slightly more autocorrelation than expected (even though the value of Moran's I is negative!).

Geary's C is equal to $(7 - 1)(42.506)/\{2 \times 18 \times 10.085) = 0.7025$, and since this is slightly less than one, this indicates slight positive spatial autocorrelation (although we are not carrying out a test of significance here).

For Tango's C_G, we will somewhat arbitrarily set the diagonal weights, w_{ii}, equal to 1 (this is convenient since it implies that we are looking at the sum of the squared differences between observed and expected values, all with equal weights to 1), and we set all of the off-diagonal elements where there is binary adjacency via the rook's case equal to 0.5. The vector $\mathbf{r}$ is simply the set of seven observed frequencies, all divided by the total of 42. The vector $\mathbf{p}$ is the vector of expected proportions, and is found by dividing the seven expected values in the table above by 42.

We find $C_G = (\mathbf{r}-\mathbf{p})'\mathbf{W}(\mathbf{r}-\mathbf{p}) = 0.0358$, with the expected value in a random pattern equal to 0.0161, and the standard deviation equal to 0.0119. (The expected value is found by summing the diagonal elements of $\mathbf{WV}_p$. This leads to a z-score of $(0.0358 - 0.0161)/0.0119 = 1.655$, and this just barely exceeds the critical value of $z = 1.645$ with a one-tailed test and $\alpha = 0.05$. The combination of squared deviations between observed and expected proportions within regions and between regions is on the margin of what we would expect, if the pattern were random. The reader may wish to attempt to follow along by writing out the 7×7 $\mathbf{W}$ matrix and finding the 7×7 $\mathbf{V}_p$ matrix, and then finding the seven diagonal elements by summing the products formed by using elements of row i of $\mathbf{W}$, and column i of $\mathbf{V}_p$.

Similarly, the spatial chi-square statistic is equal to $R = 0.2395$. Its expected value is 0.1387, and the standard deviation is 0.084, leading to a z-statistic of $(0.2395 - 0.1387)/0.084 = 1.2$. This is less than the critical value of 1.645 that would be used in a one-tailed test with $\alpha = 0.05$, and so we fail to reject the null hypothesis.

Suppose that we had set $\mathbf{W}$ equal to the identity matrix, $\mathbf{I}$. Then $R = (\mathbf{r}-\mathbf{p})^{*'} \mathbf{I} (\mathbf{r}-\mathbf{p})^{*}$, where $(\mathbf{r}-\mathbf{p})^{*}$ indicates that each of the elements in $\mathbf{r} - \mathbf{p}$ has been divided by $\sqrt{\mathbf{p}}$. This yields $R = 0.2406$, and one can note that this aspatial chi-square goodness-of-fit result reported above is a multiple of $N = 42$ times this value of R – thus $42 (0.2406) = 10.105$, which was the chi-square result of our earlier test. The expected value of R is $(1/N)$ $\text{Tr}(\mathbf{V}_p) = 0.0202$.

5.7 Global Statistics for Case–Control Data

Suppose we would like to know something about the spatial pattern of cancer cases on a map, and we're interested in whether there are clusters, over and above those that are "expected." Some parts of the map may have more cases than others only because there are more people, or perhaps more elderly people. One way to determine whether there are more cases than expected in some parts of the map is to place on the map the locations of a set of healthy controls. Those controls can be matched on important covariates, such as age. If for example we have a case location associated with a 69-year-old white male, we can also plot the location of a 69-year-old white male who does *not* have cancer. Our goal is to then compare the map of n_0 cases with the map of n_1 controls, and ask whether the map of cases is relatively more clustered (in comparison with the map of controls), and whether there are locations on the map where the likelihood of a plotted location turning out to be a case rather than a control is relatively high.

An early attempt at answering this question was made by Cuzick and Edwards (1990). Their statistic is the number of k-nearest neighbors of cases that are cases. Using their notation,

$$T_k = \sum_{i=1}^{N}\sum_{j=1}^{N} w_{ij}\delta_i\delta_j \qquad (5.33)$$

where $N = n_0 + n_1$, δ_i is equal to one if location i corresponds to a case and is equal to zero if the location corresponds to a control. In addition, w_{ij} is equal to one if location j is a k-nearest neighbor of location i, and zero otherwise. In practice then, we look at the k-nearest neighbors of a case, on a map where all cases and controls have been plotted. We count the number of those k-nearest neighbors that are cases. Then, we do this for all cases, and find the sum of those counts. The statistic is just a count of the number of cases that are k-nearest neighbors of cases (and there can be double counting; if j is a k-nearest neighbor of i, i might also be a k-nearest neighbor of j. High values of the statistic will lead to rejection of the null hypothesis.

There are different ways to think about the null hypothesis, but the commonly employed null process is one where the locations of cases and controls are fixed. Then, imagine taking all of the labels indicating whether the location is a case or a control off of the map. Finally, randomly toss those labels back onto the map. This is known as "random relabeling". The null distribution of the Cuzick–Edwards statistic can be generated by randomly relabeling the fixed locations, computing the statistic, and then repeating that many times. A histogram of the simulated statistics can then serve as the null distribution. If the Cuzick–Edwards statistic that is actually observed exceeds for example the 95th percentile of the distribution, the null hypothesis that the locations are consistent with random relabeling is rejected – cases are nearer to other cases than would be expected by chance alone.

For a map where the locations have been labeled randomly, the expected value of their statistic is

$$E[T_k] = \frac{kn_0(n_0-1)}{n_0+n_1-1} \qquad (5.34)$$

This expectation is quite easy to understand. When we are looking at a particular case, there are $n_0 + n_1 - 1$ other locations on the map, and $n_0 - 1$ of them are cases. The ratio of the latter quantity to the former is the probability that another location is a case. We multiply this ratio by k, because we are examining the k-nearest neighbors around the case, and we multiply that result by n_0 because we are carrying this out for all cases. As is often the case, the expression for the variance is more complex:

$$V[T_k] = (kn + N_s) p_1 (1 - p_1) + \{(3k^2 - k)n + N_t - 2N_s\}(p_2 - p_1^2) -$$
$$[k^2(n^2 - 3n) + N_s - N_t](p_1^2 - p_3) \tag{5.35}$$

where

$$p_j = \prod_{i=0}^{j} \frac{n_0 - i}{N - i} \tag{5.36}$$

and

$$N_s = \sum_i \sum_j w_{ij} w_{ji}; \ N_t = \sum_j \sum_i \sum_{l \neq i} w_{ij} w_{lj} \tag{5.37}$$

With the expectation and variance in hand we can carry out a z-test, making the assumption that the null distribution of T_k is normal:

$$z = \frac{T_k - E[T_k]}{\sqrt{V[T_k]}} \tag{5.38}$$

A shortcut that can be reasonable is to approximate the variance with the mean (recall that a property of the Poisson distribution, which captures randomness, is that the variance is equal to the mean).

5.7.1 A Quadrat Test for Case–Control Data

For a map with case–control data, a simple test can be carried out that tests the null hypothesis that the number of cases near controls is not greater than what would be expected by chance alone (Rogerson, 2006).

Again let n_0 be the number of cases and n_1 be the number of controls. Draw Voronoi polygons around each control. These polygons have the property that every point within the polygon is closer to its control than it is to any other control. Next, count the number of cases that are found within each of these polygons. These counts can then be compared with n_0/n_1, which is the number of cases *expected* within each polygon. More specifically, the chi-square goodness-of-fit statistic,

$$M = \frac{\sum_{i=1}^{n_1}(C_i - (n_0/n_1))^2}{(n_0/n_1)} = \frac{n_1 \sum_{i=1}^{n_1}(C_i - (n_0/n_1))^2}{n_0} \tag{5.39}$$

has a chi-square distribution, with $n_1 - 1$ degrees of freedom, when the null hypothesis that the cases are randomly distributed within the polygons surrounding controls is true. Note that when the number of cases is equal to the number of controls, as it often is in epidemiological studies, this simplifies to

$$M = \sum_{i=1}^{n_1}(C_i - 1)^2 \tag{5.40}$$

The implementation of the method is simplified further by recognizing that one does not actually need to draw the polygons. C_i is simply the number of cases that are closer to control i than they are to any other control. One can construct a matrix or table of distances from cases to all controls. Each row of the table corresponds to a case and the entries in the row are the distances to each of the controls. Next, find the minimum in each row. C_i is a count of the number of times a case has its minimum distance in column i (thus if you circled the minimum distance in each row, C_i is a count of the number of circled elements in column i). It is also important to note that the common rule-of-thumb that the chi-square test requires expectations of at least five is overly conservative. When expectations (in this case, n_0/n_1) are equal to one (and even lower – even when n_0/n_1 is ½ or ¼), the chi-square approximation is still often accurate (see Koehler and Larntz, 1980; Rogerson and Yamada, 2009: 51).

When the observed statistic exceeds its critical value, this implies that the distribution of cases across polygons is not likely to have occurred by chance alone – there are some polygons where there are significantly more cases than would have been expected.

This approach implies a particular spatial scale for the process, just as the choice of quadrat size does for the quadrat method. It is possible, for example, that clustering of cases is occurring at a larger spatial scale – in that case, we would expect to see a spatial pattern in the deviations from expectations. It is also possible that clustering of cases could be occurring *within* our polygons, and that the polygons are too large to capture the clustering.

Although perhaps a little less interesting and less applicable, one could also look at the number of controls within polygons constructed around each of the cases. In that case, we would be looking for "cool spots" rather than hot spots – the chi-square statistic would be significant when there were a large number of controls surrounding some of the cases.

Example

Suppose we have four controls at the following x- and y coordinates: (1,1); (2,4); (3,2); (3,3). There are four cases at the x- and y-coordinates: (1,2); (3,1); (2,2); (4,1). The reader may wish to plot these coordinates for a visual depiction of the example.

Squared distance from Case 1 to the controls: 1, 5, 4, 5

Case 2: 5, 10, 1, 4
Case 3: 2, 4, 1, 2
Case 4: 9, 13, 2, 5

Control 1 has one case that is closer to it than any other control (namely, Case 1), and Control 3 has three cases that are closer to it than any other control (Cases 2, 3, and 4). Therefore, $C_1 = 1$, $C_2 = 0$, $C_3 = 3$, and $C_4 = 0$, and the observed chi-square statistic is $(1-1)^2 + (0 - 1)^2 + (3 - 1)^2 (0 - 1)^2 = 6$. This is to be compared with the critical value from a chi-squared distribution with $4 - 1 = 3$ degrees of freedom. With a Type I error of 0.05, the critical value is 7.81; we therefore fail to reject the null hypothesis. The p-value is equal to 0.112.

$$\rule{3cm}{0pt}\boxed{\textbf{Example}}\rule{3cm}{0pt}$$

Cuzick–Edwards test

The distances between cases and controls are as follows:

	Case					Control				
	1	2	3	4	5	1	2	3	4	5
Case 1	-	.12	.20	.50	.60	.40	.50	.29	.61	.24
Case 2	.12	-	.34	.26	.40	.16	.34	.25	.39	.41
Case 3	.20	.34	-	.47	.62	.21	.62	.31	.17	.26
Case 4	.50	.26	.47	-	.16	.34	.24	.70	.29	.19
Case 5	.60	.40	.62	.16	-	.23	.70	.61	.38	.49

Suppose that we have 5 cases and 5 controls.

For each case, we're going to look at the k-nearest neighbors, and ask how many of them are cases. We know how far the cases are from the other points. Let's look at the first row of the matrix of case–control distances. Case 1 has two cases as its closest neighbors. Case 2 has one case among its two closest neighbors. Among the two closest neighbors, cases 3, 4, and 5 each have one case. The Cuzick–Edwards statistic is 6 – that is, the number of case–case pairs that are 2-nearest neighbors of one another. We're not assuming that the places are randomly situated on the map. Instead, we are assuming for our null process that the fixed locations have been randomly labeled with case–control status. The expectation of the statistic is found from Equation 5.28:

$$2(5)(5 - 1)/(5 + 5 - 1) = 40/9 = 4.44$$

We have observed a few more case–case pairs than we would have expected. The intuition behind the expected value is as follows: four of the other nine locations on the map are cases – this is the chance that a nearest neighbor is a case. We multiply by the number of cases (since we are doing this for each case) and multiply by 2, since we are looking at the closest two neighbors in this case. We can do an approximate test of statistical significance by forming the z-statistic:

$$z = (6 - 4.44) / \sqrt{4.44} = 0.74$$

and this is not high enough to reject the null hypothesis of random labeling.

(**Example**)

Modified Cuzick–Edwards/quadrat test

We're going to look at each control, and carve out an area around each. These areas have the characteristic that every point within the area is closer to that control than it is to any other control. We'll compare what is observed with what is expected. Since there are five cases spread across the five areas created around the controls, we will expect $n_0/n_1 = 5/5 =$ one case in each of these areas.

For each case, we want to know where the closest control is. For each row in the matrix of case–control distances (where cases are represented by the rows, and controls are represented by the columns), we find the minimum value. In the first row, note that Case 1 is closer to Control 1 than it is to any other control. Repeat this for all cases (e.g., Case 2 is closer to Control 1 than it is to any other control. Case 3 is closest to Control 4). Case 4 is closest to Control 5, and Case 5 is closest to Control 1. The chi-square statistic then is found by adding up the number of cases that have been associated with (i.e., are within the area of) each control. Controls 1 and 5 have two cases associated with them, Control 4 has one case, and Controls 2 and 3 both have no cases associated with them:

$$(2 - 1)^2/1 + (0 - 1)^2/1 + (0 - 1)^2 /1 + (1 - 1)^2 /1 + (2 - 1)^2/1 = 4 + 1 + 1 = 4$$

The next step is to compare this with the critical value of chi-square that has degrees of freedom equal to the number of categories (i.e., the number of controls), minus one. With $\alpha = 0.05$ and four degrees of freedom, this critical value is equal to 9.49. Since the observed value is less than this, we fail to reject the null hypothesis that the cases are randomly distributed across the control areas.

Exercises

1. Repeat the exercise above for COVID-19, this time for the following counties in New York State:

County	Cases (as of 3/28/2020)	Population
1 Steuben	11	95796
2 Seneca	0	34300
3 Ontario	18	109864
4 Yates	0	24841

(Continued)

5 Schuyler	0	17912
6 Chemung	15	84254
7 Livingston	3	63227
	Total 47	

a. Find a New York State map, locate the counties, and find the weight matrix for counties sharing a border.
b. Find expected numbers of cases by multiplying 47 by each county's share of total population.
c. Carry out the chi-square, aspatial goodness-of-fit test.
d. Find z-scores for each county (observed – expected)/sqrt(expected)
e. Find Moran's I and Geary's C for the set of z-scores.
f. Now set the diagonal elements of the weight matrix equal to 1, and the off-diagonal elements that correspond to adjacent regions equal to 0.5. Find Tango's statistic, and its expectation. Find the spatial chi-square statistic and its expectation.

2. A 6 x 6 set of quadrats is placed on top of a study area containing 80 points. The number of quadrats containing a given number of points is as follows:

Number of points	Frequency (i.e., number of quadrats containing that many points)
0	9
1	12
2	12
3	6
4	4
5	2

Use the quadrat test to test the null hypothesis that the pattern is random. State your conclusion and give the p-value.

3. The following data summarize the number of cases of crime observed and expected in a set of five regions:

Region	Observed	Expected
A	11	7
B	12	6
C	11	10
D	8	13
E	8	14

Use the chi-square goodness-of-fit test to test the null hypothesis that the observed number of cases is not inconsistent with expectations. State your conclusion and give the p-value.

4. Construct a six-region system by using six quadrats – arranged in two rows of three cells each (i.e., three squares, side by side, and then three squares under those). Use rook's case binary adjacency for the weights.

   Label the regions 1, 2, 3 in the first row, going from left to right. Label them 4, 5, and 6 in the second row, again going from left to right. Use the following regional values, for regions 1-6: 7, 4, 8, 2, 3, 6. Find Moran's I and Geary's C.

Hint: Each term appears twice in the numerator. If Region 1 is connected to Region 2, then Region 2 is connected to Region 1. You can find the product for just one instance for each pair; then double the result when you are done.

Further reading

Good introductory coverage of global methods such as quadrat methods and Moran's I is provided by Dale and Fortin (2014). Cuzick and Edwards (1990) is the standard reference for detecting geographic patterns in case–control data.

Online resources

Visit **https://study.sagepub.com/rogersonspatialstatistics** for resources which supplement the material presented in this chapter. Students can find datasets and additional exercises, which can be used to practice the techniques covered in this chapter, as well as a chapter introduction video and video demonstration of one of the most commonly used global spatial statistics, Global Moran's I. The topic-software chart also describes the software packages that can be used to implement key techniques.

Lecturers and instructors can find exercises and conceptual prompts to facilitate classroom discussion and practice.

6

Local Tests

6.1 Introduction

In many cases, we have a predefined location on the map, and there is interest in knowing whether incidence of, for example, crime or disease is raised around that particular location. Are there more cases than one would expect (say, based upon the size and possibly characteristics of the population, such as age) in the vicinity of that location?

We will take our global statistics and disaggregate them into component parts that are associated with specific locations. The global statistic can be thought of as the sum of these components, and these components are the local statistics that correspond to specific subregions. Local tests are sometimes referred to as focused tests.

Suppose that you carried out a global test, and it was significant; the null hypothesis of spatial randomness was rejected. You would want to know *why* it was significant, and what regions contributed to that finding. There are two quite different scenarios. One is where there are many regions where the number of observations just barely exceed expectations. No local test would have shown raised incidence in any single region, but the cumulative effect of the small excess observations across many regions was sufficient to reject the global hypothesis. Another possibility is that there is just one or a very small number of regions where observations are substantially greater than expectations. Here the global significance may be attributed to those regions; without those observations in those regions, the global statistic would not have been significant.

Now suppose that the global test was not significant. It is still possible that individual local tests could be significant. Such outliers occur when there are local pockets of significant excess risk that are unlikely to have occurred by chance alone, but they occur within the context of many other regions. The excess risk in one or a small number of locations does not have to be of such magnitude that it leads to a significant global statistic.

6.2 Local Quadrat Test

The conditional quadrat test discussed in Section 5.1 was a global test based upon a comparison of observations and expectations. This comparison was carried out cell by cell – the function of the difference for each cell was added across cells (that is, the deviations between observed and expected values were first squared, and then the result was divided by the expectation). If we are interested in a particular cell i, a simple local test for that cell is

$$\chi_i^2 = \frac{(O_i - E_i)^2}{E_i} \tag{6.1}$$

This has a chi-square distribution, with one degree of freedom, if the null hypothesis of no raised incidence is true. The critical value for $\alpha = 0.05$ is equal to 3.84. An alternative and equivalent version of this is to form the z-score

$$z = \frac{(O_i - E_i)}{\sqrt{E_i}} \tag{6.2}$$

Note that when the usual critical value of $z = 1.96$ is squared, the result is 3.84, the critical value of the chi-squared variable with one degree of freedom.

For example, if we expected 20 cases in a region and we observed 25, how unusual would this be? The z-score is $(25 - 20)/\sqrt{20} = 1.12$, and this is less than, for example, the critical value of $z = 1.645$ for a one-sided hypothesis test with $\alpha = 0.05$; there is not enough evidence to reject the null hypothesis that observations are consistent with expectations.

A couple of points can be made here – one is that there is a discreteness to the statistic, since observed counts only take on integer values. Thus under the null hypothesis, the distribution of the statistic cannot be normal, and in fact it is not even continuous. The binomial and Poisson distributions are associated with better descriptions of the underlying process. Still, the normal approximation works reasonably well.

Also, the distribution of the statistic is closer to normal under the unconditional null process than it is under the conditional one. The conditional distribution is one we discussed in the previous chapter – N observations are randomly tossed onto a map of m regions. Here there is a covariation between the observations among cells – if there are more observations in one cell, there are likely to be less in another (because of the fixed total number of observations). The unconditional process treats each cell as an observation from a Poisson distribution having a mean of N/m – and hence the result is not conditional on any total number of observations. The z-test described above approximates the Poisson distribution (and this can be seen in part by recognizing that the square root of the number of expected cases in the denominator corresponds to the standard deviation of a Poisson variable).

6.3 Stone's Test

Stone's (1988) statistic is the maximum ratio of cumulative observed to cumulative expected cases, when cases have been arranged in order of increasing distance around the focal location of interest. See Table 6.1, where four zones around the point of interest have been arranged according to increasing distance. There are 50 cases; 10 of them are observed in the first zone, when only 5 were expected. The ratio is equal to 2, and since this is the highest of all of the ratios, it serves as the observed value of Stone's statistic.

Table 6.1   Hypothetical data for Stone's test

Cumulative observed	Cumulative expected	Ratio
10	5	2
20	16	1.25
35	30	1.17
50	50	1

One can assess the statistical significance of the statistic by simulating the null hypothesis of randomness. To do this, take the observed number of cases and allocate them to regions according to expectations, and then find the maximum ratio of observed to expected cases, across all regions. (As a reminder, we are setting aside the issue of where the expected number of cases comes from – typically these might be based on population, perhaps adjusted for age and sex; they might also come from a more sophisticated model that predicts the observed number of cases using covariates.) This simulation could be repeated, for example, 1000 times, and the 1000 maximum ratios could be ranked from lowest (rank 1) to highest (rank 1000). This is termed the null distribution. The 950th simulated value on the ranked list would then serve as the critical value for a one-sided test using $\alpha = 0.05$.

One issue with this method is the fact that the maximum ratio can be quite unstable. If the cumulative expected value is small, then the ratio can be quite large. For example, if only one case is expected in the closest region, and two are observed, the ratio of 2 implies that there are twice as many cases as are expected – but this might not be important, since only one case was expected. An alternative that we will explore next is a statistic based upon the *differences* between cumulative expectations and observations.

6.4 Kolmogorov–Smirnov Test

An alternative to Stone's test that avoids the instabilities that accompany small denominators is to use the maximum *difference* between cumulative observed and cumulative expected values – this is identical to what is known as the Kolmogorov–Smirnov statistic, for data that have been ordered.

Conover (1972) gave a recursive algorithmic approach for assessing the significance of this statistic. Rogerson (2012) discussed its use in the context of local spatial testing with case–control data.

The statistic is

$$d^+ = \frac{1}{C} \max_{k=1,\ldots,K-1} \sum_{i=1}^{k}(C_i - E_i) \tag{6.3}$$

where there are K zones arranged in order around the focal location, C_i and E_i are the observed and expected counts in zone i, and C is the total number of cases. We seek the p-value, $\Pr(D^+ \geq d^+)$, where d^+ is the observed value of the statistic. The steps are as follows:

1. Let $H_0 = 0$, and $H_k = (1/C)\ \sum_{i=1}^{k} E_i$ for $k = 1, 2, \ldots, K$.

2. Set $f_j = 1 - d^+ - \dfrac{j}{C}$ if this value corresponds to one of the H_k; if not, set it equal to the largest among the values of H_k that are less than $1 - d^+ - \dfrac{j}{C}$.

3. Let $e_0 = 1$, and for $k \geq 1$,

$$e_k = 1 - \sum_{j=0}^{k-1} \binom{k}{j} f_j^{k-j} e_j$$

continuing for all values of k where $f_k > 0$.

4.   The p-value is given by

$$\Pr\left(D^+ \geq d^+\right) \; = \sum_{j=0}^{[C(1-d^+)]} \binom{C}{j} f_j^{n-j} e_j$$

where the square brackets imply that the integer part of the quantity is to be taken.

An illustration is given in Section 6.9.

6.5 Score Statistic

While local Moran's I (Section 6.8) is perhaps the most widely used local statistic in geography, the score statistic is used often in epidemiological applications – different fields have different tendencies to use particular approaches.

The score statistic is

$$C_i = N \sum_{j=1}^{m} w_{ij} \left(r_j - p_j\right) \tag{6.4}$$

where r_j and p_j are the observed and expected proportions of cases falling in region j, respectively, and N is the total number of cases. The weight w_{ij} is a measure of the connectedness of region i to region j; typically, these weights will be higher near the location of interest and lower far away from it. Again, we wish to test the null hypothesis that there is no raised incidence around location i. If C_i is high and positive, we will reject the null hypothesis, in favor of the alternative that observations are unlikely to have exceeded expectations in the vicinity of the focal location by chance alone.

In this case, unlike Moran's I, w_{ii} is not chosen to be zero; rather it is usually chosen to be higher than w_{ij} when j is not equal to i. This is consistent with the idea that under the alternative hypothesis, observations are likely to exceed expectations to a greater degree in region i than in other regions, j. Note that Moran's I focuses upon the relationships between pairs of regions; the score statistic focuses upon individual regions that are in the vicinity of the location of interest.

As always, we are interested in the distribution of the statistic when the null hypothesis is true. The expected value is equal to zero; it is easy to see from Equation 6.4 that C_i will be equal to zero if observations are equal to expectations in each region. The variance is

$$V[C_i] = N \left\{ \sum_{j=1}^{m} w_{ij}^2 p_j - \left(\sum_{j=1}^{m} w_{ij} p_j \right)^2 \right\} \qquad (6.5)$$

From this we can carry out a z-test

$$z_i = \frac{C_i}{\sqrt{V[C_i]}} \qquad (6.6)$$

One important point is that this statistic is a uniformly most powerful statistic – if incidence is actually raised around the location of interest, according to a risk profile (which can be thought of as a graph of risk versus distance) consistent with the choice of weights, no other statistic will be as good at finding the raised incidence.

The matched-filter theorem is relevant here. This theorem implies that it is important to give careful consideration to the choice of weights. If actual risk is defined by some function (e.g., some particular decline in risk with distance as one goes away from location i), the statistical power associated with the score statistic will be highest when weights are chosen to match the actual change in risk. If we guess that risk declines quickly with distance (i.e., there is a cluster small in its spatial extent) but that is not the case, the statistical power – the ability to find the cluster – is reduced. Likewise, if we guess that risk declines slowly with distance, but it actually declines quite rapidly, we are assuming that there is a cluster that is larger in spatial extent than is actually there. Again, the power is compromised. The matched-filter theorem states that we have the best chance of detecting the cluster if we choose weights that match the way in which actual risk changes with distance. Of course the big problem and major challenge is that we don't *know* the actual risk profile – that is why we are doing the test! One approach would be to try different sets of weights, but then of course one would have to worry about multiple testing. It isn't right to try lots of different weight functions, and simply report the one with the lowest *p*-value! But it does make sense to at least look at different weights, especially if the analysis is being conducted more in an exploratory mode (rather than in a confirmatory mode), where one is interested in the sensitivity of results to different specifications of the weights.

We note here that Tango's statistic is the sum of the squared local score statistics; that is,

$$C_G = \sum_{i=1}^{m} C_i^2 \qquad (6.7)$$

where, if the weights are given the notation w_{ij} for the local score statistic, then the corresponding global Tango statistic derived by summing the squared local score statistics would have weights equal to $a_{ij} = \sum_k w_{ki} w_{kj}$.

6.6 Modified Local Score Statistic

A modified local score statistic is

$$R_i = N\sum_{j=1}^{m} \frac{w_{ij}\left(r_j - p_j\right)}{\sqrt{p_j}} \tag{6.8}$$

This is simply the local score statistic from the previous section, with the weights defined as $w_{ij}^* = w_{ij}/\sqrt{p_j}$. The sum of these squared local statistics, namely $\sum_{i=1}^{m} R_i^2$, is equal to the global spatial chi-square statistic, R, where the weights in the global statistic are equal to $a_{ij} = \sum_k w_{ki}w_{kj}/\sqrt{p_i p_j}$. For the global statistics, recall that we found that Tango's statistic had higher power than the spatial chi-square statistic when the increased risk occurred in a place with high population. The spatial chi-square statistic had relatively higher power when the increased risk occurred in a small place. And yet at the local level, the score statistic is uniformly most powerful – it generally does a little better at rejecting the null hypothesis when it is false, in comparison with R_i.

An alternative form of local statistic might be considered so that the sum of these local statistics added to either Tango's global statistic or the global spatial chi-square statistic. Thus we could have

$$\left(r_i - p_i\right)\sum_j w_{ij}\left(r_j - p_j\right) \tag{6.9}$$

as a local Tango statistic, and with modified weights (dividing by $\sqrt{p_i p_j}$) we would have a local spatial chi-square statistic. These will not be pursued further here. The existing relationships between (a) the local score statistic and Tango's statistic, (b) the modified local score statistic and the global spatial chi-square statistic, and (c) the local quadrat test and the aspatial chi-square statistic are attractive ways to define local statistics.

6.7 Getis–Ord Statistic

The Getis–Ord statistic and local Moran's I (Section 6.8) are the two most commonly used local statistics in geography. Again, our objective is to find spatial clustering around a point. Local statistics are weighted sums of observations around the locality of interest. Getis and Ord (1992) and Ord and Getis (1995) set out their statistics in two papers. In the first paper they described their statistic as the proportion of a variable that lies within a specified distance of a locality. They used a 0–1 definition for their weights:

$$G_i = \frac{\sum_{j=1}^{m} w_{ij}\left(d\right) x_i}{\sum_i x_i} \tag{6.10}$$

where m is the number of regions and $w_{ij}(d)$ is equal to 1 if region j is within a distance d of region i. They derived the expectation and variance of the statistic to allow testing of the null hypothesis that no excess (or deficient) amount of x was in the neighborhood of the locality. In their second paper, they generalized and simplified their original contribution in a number of ways. They allowed a general definition for the weights, they standardized their statistic to have a mean of 0 and a variance of 1 so that the null distribution would have a standard normal distribution, and they defined two separate versions – G_i does not include the focal region i in the sum, and G_i^* does. Thus

$$G_i^* = \frac{\sum_j w_{ij} x_j - W_i^* \bar{x}}{s\left[\dfrac{mS_{1i}^* - \left(W_i^*\right)^2}{m-1}\right]^{1/2}} \tag{6.11}$$

where m is the number of regions, s is the standard deviation of the data, W_i^* is the sum of the weights $(\sum_j w_{ij})$, and S_{1i}^* is the sum of the squared weights $(\sum_j w_{ij}^2)$.

Since the statistic has, approximately, a standard normal distribution (with mean 0 and variance 1), it can be treated as a standard normal variate (i.e., a z-score).

Essentially, when the weights are 0–1, the statistic compares the values of x within a neighborhood of the focal location with the values of x outside of the neighborhood. Although it has not been widely noted, in fact their statistic in that case is identical to a two-sample t-test.

6.8 Local Moran Statistic

The local Moran statistic is a commonly used approach for examining the hypothesis that incidence is raised around a location of interest (i). This statistic is based on a component of the global Moran statistic. In particular one could define the quantity

$$\frac{n(x_i - \bar{x})\sum_{j=1}^n w_{ij}(x_j - \bar{x})}{\left(\sum_{i=1}^n \sum_{j=1}^n w_{ij}\right)\sum_{i=1}^n (x_i - \bar{x})^2} \tag{6.12}$$

The sum of the local Moran statistics, if defined this way and summed over all regions, would be equal to the global statistic. The actual definition of local Moran's I as defined by Anselin (1995) in his original article, however, is similar to this, and is a multiple of the quantity in Equation 6.12:

$$I_i = (x_i - \bar{x})\sum_{j=1}^n w_{ij}(x_j - \bar{x}) = z_i \sum_j w_{ij} z_j \tag{6.13}$$

where the z's here are simply the x's with the mean subtracted from them. Under the null hypothesis that the observed regional pattern is a random permutation of all possible rearrangements of regional values, the expected value is equal to $E[I_i] = -\Sigma_j w_{ij} / (n-1)$, where n is the number of regions. The variance is

$$V[I] = \frac{(n-b_2)\Sigma_j w_{ij}^2}{(n-1)} - \frac{(2b_2-n)\Sigma_{k\neq i}\Sigma_{h\neq i}w_{ik}w_{ih}}{(n-1)(n-2)} \qquad (6.14)$$

where

$$b_2 = \frac{\Sigma_i \dfrac{z_i^4}{n}}{\left(\Sigma_i \dfrac{z_i^2}{n}\right)^2} = n\frac{\Sigma_i z_i^4}{\left(\Sigma_i z_i^2\right)^2} \qquad (6.15)$$

Assuming that the statistic is approximately normally distributed for large samples, the expressions for the expectation and variance given above allow for the usual z-test.

It is common to map local Moran's I, showing regions of high values that are surrounded by other high values and also showing regions of low values that are surrounded by other low values. If global Moran's I is significant, these subareas of a map depict the locations that are contributing to the global significance. Even when Moran's I is not significant, there still may be pockets (outliers) where the local statistic is significant. Such maps also depict areas where low (high) values are surrounded by high (low) values; these have values for the local Moran statistic that are negative, and they contribute to negative global spatial autocorrelation.

6.9 Illustrations

Suppose that we have information on the number of cases that are observed and expected (with the latter perhaps derived from a separate analysis of population and relevant covariates such as age) in each of four regions that are ordered in terms of increasing distance around a location of interest (Table 6.2). The table also depicts weights that will be used in Section 6.9.3.

There are a total of 50 cases in the study area. Note that the columns displaying the observed and expected proportions sum to 1. We wish to test the null hypothesis that there is no raised incidence around the location of interest.

Table 6.2 Illustrative data for local tests

Observed	Expected	Cumulative observed	Cumulative expected	Observed proportion (r)	Expected proportion (p)	Weight
16	10	16	10	0.32	0.20	1
12	10	28	20	0.24	0.20	0.5
14	20	42	40	0.28	0.40	0.3
8	10	50	50	0.16	0.20	0.2

6.9.1 Stone's Test

To carry out Stone's test, we first find the test statistic as the maximum ratio of cumulative observed to cumulative expected frequencies, as zones are examined sequentially. This is equal to $16/10 = 1.6$. Next, we need to determine how unusual such a ratio is, if the null hypothesis is indeed true. To simulate the null hypothesis, 50 random numbers were chosen from a uniform distribution on the interval $(0,1)$. Each of the numbers was assigned to a region based upon the following: $(0–0.2) \rightarrow$ Region 1; $(0.2–0.4) \rightarrow$ Region 2; $(0.4–0.8) \rightarrow$ Region 3; and $(0.8–1.0) \rightarrow$ Region 4. These ranges and assignments are in line with the fact that the proportion of cases expected to fall in each region is $\{0.2, 0.2, 0.4, \text{ and } 0.2\}$ and this in turn is based upon the expected number in Table 6.2, relative to the total of 50 cases $\{10/50, 10/50, 20/50, \text{and } 10/50\}$. Once the 50 cases were assigned to regions, Stone's statistic was calculated. This was repeated 100,000 times, and the resulting statistics were ordered from lowest to highest. The $95,000$th statistic on the ordered list was 1.60, and this serves as the critical value for a one-sided test with $\alpha = 0.05$. The observed statistic is identical, and so the p-value is 0.05; the decision is right at the border between rejecting and failing to reject the null hypothesis.

6.9.2 Kolmogorov–Smirnov Test

Following the steps outlined in Section 6.4 along with the data in Table 6.2, the H-values are based upon the cumulative expected proportion of cases, by ordered region: $Hi = \{0, 16/50 = 0.32, 0.56, 0.84, 1\}$; $i = 0, ..., 4$. The f's are $f_0 = 0$, $f_i = 0.56$ for $i = 1, ..., 14$, and $f_i = 0.32$ for $i = 15, ..., 26$. For $i > 26$, $f_i = 0$. The values of e are found via recursion and are equal to $\{0.16, 0.1152, 0.0632, 0.0313, ...\}$, for $i = 1,...$ (and $e_0 = 0$). Although these values are found for all i up to 26 (which is the highest value of i for which $f_i > 0$), they decline rapidly, and are less than 0.0001 for $i > 11$. Combining these quantities using the four steps in Section 6.4, we find the p-value to be 0.0252, and therefore the null hypothesis of no raised incidence around the focal location is rejected.

6.9.3 Score Statistic

The score statistic is based upon a weighted sum of the differences between observed and expected values. We will decrease the weight for each region as one moves away from the location of interest; these weights are chosen prior to the analysis, and, for this illustration, are arbitrary (recall that in practice we want to choose them in a way that matches the profile of risk as it changes across regions) and given in Table 6.2. The statistic is calculated as follows:

$$C_i = N\sum_j w_{ij}\left(r_j - p_j\right)$$
$$= 50\{1(0.32-0.20)+0.5(0.24-0.20)+0.3(0.28-0.40)+0.2(0.16-0.20)\} \quad (6.16)$$
$$= 50(0.096) = 4.8$$

The expectation of the score statistic is 0; since our value is positive, this is consistent with observations that are higher than expected in the vicinity of the location of interest. The variance is found as

$$V[C_i] = N\left[\sum_j w_{ij}^2 p_j - \left(\sum_j w_{ij} p_j\right)^2\right] = 50[0.0824] = 4.12 \tag{6.17}$$

The z-score associated with the statistic is $(4.8 - 0)/\sqrt{4.12} = 2.36$, and this is greater than the critical value of 1.645 that would be used with a one-sided test with $\alpha = 0.05$ and so the null hypothesis is rejected.

6.9.4 Local Version of the Spatial Chi-Square Statistic

By modifying the weights used in the score statistic (by dividing each weight w_{ij} by the square root of the expectation, p_j), we have the local version of the spatial chi-square statistic:

$$R_i = N\sum_j w_{ij}(r_j - p_i)/\sqrt{p_i}$$
$$= 50\left\{\frac{1(0.32-0.20)}{\sqrt{0.2}} + \frac{0.5(0.24-0.20)}{\sqrt{0.2}} + \frac{0.3(0.28-0.40)}{\sqrt{0.4}} + \frac{0.2(0.16-0.20)}{\sqrt{0.2}}\right\} \tag{6.18}$$
$$= 50(0.238) = 11.9$$

Like the score statistic, this has an expectation of 0. Its variance is found from

$$N\left[\sum_j w_{ij}^{2*} p_j - \left(\sum_j (w_{ij}^*)^2 p_j\right)^2\right] \tag{6.19}$$

using the modified weights ($w_{ij}^* = w_{ij}/\sqrt{p_j}$); this is equal to $V[R_i] = 23.875$. Thus the z-score is $(11.9 - 0)/\sqrt{23.88} = 2.44$. This result is similar to that found for the score statistic, and again the null hypothesis is rejected.

6.9.5 Getis–Ord G_i Statistic

The Getis–Ord statistic tests whether the values near the location of interest are different than the values far from the location. Their original statistic used binary weights, but in their subsequent paper they relaxed this restriction. Here we will test whether the z-scores associated with individual regions are higher near the focal location than they are away from the focal location. We will arbitrarily examine the case for weights of {1, 1, 0, 0} for the four respective regions; this is equivalent to testing whether the z-scores for the two regions near the focal point are different from the scores for the two regions that are the farthest away from it. The z-scores are found from $z = $ (observed − expected)/$\sqrt{\text{expected}}$.

For the four regions, we find $z = \{1.9, 0.632, -1.34, -0.632\}$. The z's are positive in the first two regions, corresponding to the fact that observations exceed expectations. The Getis–Ord statistic is

$$G_i^* = \frac{\Sigma_j w_{ij} x_j - W_i^* \bar{x}}{s\left[\left(mS_{1i}^* - \left(W_i^*\right)^2\right)/(m-1)\right]^{1/2}} = \frac{(1.9+0.632)-2(0.1389)}{1.429\left[\left(8-2^2\right)/(4-1)\right]^{1/2}} = 1.37 \tag{6.20}$$

where $m = 4$ regions, $W_i^* = 2$, and $S_{1i}^* = 2$. The observed statistic is less than the critical value of 1.645, and so we fail to reject the null hypothesis.

6.9.6 Local Moran's I

Suppose that we have the map shown in Figure 6.1a. There are nine regions, and the x-values are given for each of these regions. The mean of these values is 4, and the map of values with mean subtracted is shown in Figure 6.1b.

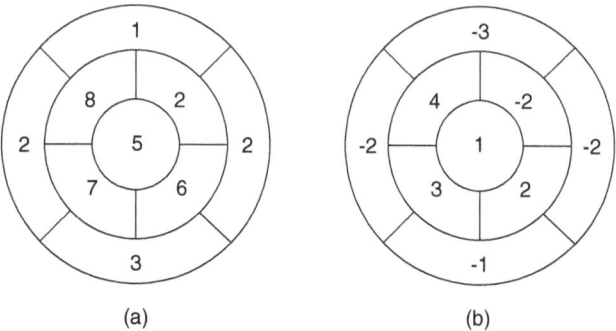

(a)   (b)

Figure 6.1   A nine-region system: (a) hypothetical x-values; (b) deviations from the mean

To find the local Moran statistic for the central region in the figure, we first find the weights. The weights, using binary connectivity for adjacent regions, are equal to 1 for the four surrounding regions, and 0 for the other regions. We then have:

$$I_i = z_i \Sigma_j w_{ij} z_j = 1\left[1(2)+1(3)+1(4)+1(-2)\right] = 7 \tag{6.21}$$

The expected value of the local Moran, using binary adjacency for the weight definition, is $-4/(9-1) = -0.5$. For the variance we have the quantities

$$b_2 = \frac{\Sigma_i \frac{z_i^4}{n}}{\left(\Sigma_i \frac{z_i^2}{n}\right)^2} = \frac{\left(\frac{484}{9}\right)}{\left(\frac{52}{9}\right)^2} = 1.61 \tag{6.22}$$

$$V[I] = \frac{(n-b_2)\sum_j w_{ij}^2}{(n-1)} + \frac{(2b_2-n)\sum_k \sum_h w_{ik}w_{ih}}{(n-1)(n-2)} = \frac{(9-1.61)(4)}{9-1} + \frac{(2(1.61)-9)(16)}{(9-1)(9-2)} \qquad (6.23)$$

$$= 2.044$$

The "16" in the numerator of this equation comes from the fact that there are four regions surrounding i (and serving as the cases where w_{ik} and w_{ih} = 1), and these form 4 x 4 =16 pairs including those where $k = h$. The z-score associated with the null hypothesis is $z = \{7 - (-0.5)\}/ \sqrt{2.044} = 5.25$ and so the null hypothesis of no local spatial association is convincingly rejected.

6.10 Local Statistics for Case–Control Data

Suppose that we have counts of the numbers of cases and controls in zones arranged around a location of interest in order of increasing distance. In the example in Table 6.3, we have 70 cases and 70 controls. We see from the table that there are indeed quite a few cases relative to controls near the site of interest. How do we test the null hypothesis that there is no excess risk of cases near the focal location?

Table 6.3 Hypothetical case–control data, arranged by zone

Zone	Cases	Controls	Row sum
1	10	6	16
2	12	5	17
3	20	20	40
4	20	25	45
5	8	14	22
Total	70	70	140

With the cumulative chi-square statistic, one looks at separate 2 × 2 chi-square tables based on the observed data, divided into all possible definitions of "near" and "far." These tables are compared with their respective "expected" tables by computing a chi-square statistic that compares the two. The expected tables are consistent with the null hypothesis that the relative prevalence of cases does not vary with distance from the location of interest.

We begin by constructing a series of 2 × 2 tables. The first of these compares zone 1 with the remaining zones:

Zone	Cases	Controls	Row sum
1	10	6	16
2–5	60	64	124

Next, find the other three 2 × 2 tables

Zone	Cases	Controls	Row sum
1–2	22	11	33
3–5	48	59	137

and

Zone	Cases	Controls	Row sum
1–3	42	31	73
4–5	28	39	67

and finally,

Zone	Cases	Controls	Row sum
1–4	62	56	118
5	8	14	22

Each one of these observed 2 × 2 tables can be compared with the table that would be expected if there were no interaction between rows (regions, divided into "near" and "far") and columns (case–control status). The entry in each expected table is equal to the row sum times the column sum, and then this product is divided by the overall total. For example, the entry in the upper left of the first table of expected values is equal to $(16)(70)/140 = 8$. The other entries can then be determined using the facts that the row and column sums of the expected table must equal those of the observed table. Thus the 2 × 2 table of expected values for comparing region 1 with regions 2–5 is

Zone	Cases	Controls	Row sum
1	8	8	16
2–5	62	62	124
Total	70	70	140

Next, the chi-square goodness-of-fit statistic is found using the four observed and expected values from each table via $\chi^2 = \sum_{i=1}^{4}(obs_i - exp_i)^2 / exp_i$. This is done for each of the four sets of tables (corresponding to the four different spatial definitions). Cumulating the successive chi-squares associated with the four sets of observed and expected tables, we have

$$\chi^2_{cumulative} = T = \sum_{k=1}^{K-1} \chi_k^2 = 1.13 + 4.80 + 3.46 + 1.94 = 11.33 \tag{6.24}$$

where K is the number of zones, and the critical value for a chi-square random variable with the one degree of freedom associated with a 2 × 2 table is 3.84. Thus we see that if we were simply comparing the first two zones (which have 22 cases and 11 controls) with the other three zones, the observed-value of chi-square is 4.80, and the null hypothesis would be rejected. But we must account for the fact that we've done four tests – we can't simply look at them all and take the best of the results.

The distribution of T/d when the null hypothesis is true is approximately chi-squared, with v degrees of freedom, where $v = (K-1)/d$ and

$$d = 1 + \frac{2}{K-1}\left\{\frac{\lambda_1}{\lambda_2} + \frac{\lambda_1+\lambda_2}{\lambda_3} + \ldots + \frac{\lambda_1+\lambda_2+\ldots+\lambda_{K-2}}{\lambda_{K-1}}\right\} \tag{6.25}$$

and

$$\lambda_i = \frac{N_1+N_2+\ldots+N_i}{N_{i+1}+\ldots+N_K} \tag{6.26}$$

where N_i refers to the sum of row i.

With our illustrative example then, $\lambda_1 = 16/124 = 0.129$; $\lambda_2 = (16+17)/(40+45+22) = 33/107 = 0.308$; $\lambda_3 = 73/67 = 1.090$; and $\lambda_4 = 118/22 = 5.36$. We have

$$d = 1 + (2/4)\{0.129/0.308 + 0.437/1.09 + 1.527/5.36\} = 1.552 \tag{6.27}$$

The degrees of freedom are equal to $v = (K-1)/d = (5-1)/1.552 = 2.577$. Using an online app for a chi-square distribution (which is more useful than the chi-square tables in the back of statistics books, since they allow for fractional degrees of freedom), we find that the critical value of chi-square, using $\alpha = 0.05$, is 7.066. Since our observed value of $T/d = 11.33/1.552 = 7.300$ is greater than this, we reject the null hypothesis. The p-value, found from an online app for the chi-square distribution, is about 0.045.

There are a few approximations that are well worth mentioning, since they avoid much of the burden of calculation above.

An approximation for the degrees of freedom is

$$v \approx \frac{(K-1)^2}{\left(2K^2\sum_{i=1}^{K}i^{-2}\right)-\left(3K^2-2K+1\right)} \tag{6.28}$$

In our example, with $K = 5$, this works out to be $v \approx 2.23$. Note that with this expression, calculation of the λ's is no longer necessary. Since T/d has a chi-square distribution with v degrees of freedom, and $d = (K-1)/v$, we see that $vT/(K-1)$ has a chi-square distribution with v degrees of freedom.

If K is large (over about 20 or 30), the degrees of freedom approach the constant 3.45. In that case, T/d has a chi-square distribution with 3.45 degrees of freedom, and $3.45T/(K-1)$ has a chi-square distribution with 3.45 degrees of freedom. (Usually we think of chi-square distributions as having integer degrees of freedom, but this does not need to be the case) If the null hypothesis is rejected, the implication is that the distribution of cases and controls across zones is unlikely to have occurred by chance – there is an unusually large difference between at least one of the pairs of observed and expected tables. The test itself does not indicate *which* definition of "near" and "far" has "caused" the rejection; that is, the spatial scale of the deviation from randomness is not an outcome of the test.

Finally, if detailed locations for cases and controls is known, it is not even necessary to aggregate cases and controls into zones. Each row of the table can contain information on just one location – and it is either a case or a control, with a "1" in one column and a "0" in the other (so that the row sum is equal to one). Although this pushes the recommended minimum cell frequency for the chi-square test to its limit (the common rule of a minimum frequency of 5 is far too conservative, and a minimum frequency of 1 is still usually quite reasonable), the approximation is still not bad.

6.10.1 Maximum Chi-Square Test

There is also a maximum chi-square test – instead of cumulating the chi-square values associated with the 2 × 2 tables, we can use the maximum of the chi-square statistics as our statistic. The spatial scale associated with this maximum chi-square statistic is suggested automatically – the definitions of near and far that are associated with that particular table that yields the highest chi-square statistic also constitute estimates of the spatial extent of raised incidence.

How do we assess the statistical significance of the observed maximum chi-square statistic (denoted d)?

Here we will follow the heuristic approach to finding p-values described in Boulesteix (2006); this in turn is related to earlier work of Miller and Siegmund (1982) and Koziol (1991). Application to local statistics is discussed in Rogerson (2012). Using the notation there, let N_1 and N_2 be the number of cases and controls, respectively, with $N = N_1 + N_2$. Then

1. Find the functions

$$\text{lower}_a(x) = \frac{N_2 x}{N} - \frac{N_1 N_2}{N} \sqrt[a]{\frac{x}{N}\left(1 - \frac{x}{N}\right)\left(\frac{1}{N_1} + \frac{1}{N_2}\right)}$$

$$\text{upper}_a(x) = \frac{N_2 x}{N} + \frac{N_1 N_2}{N} \sqrt[a]{\frac{x}{N}\left(1 - \frac{x}{N}\right)\left(\frac{1}{N_1} + \frac{1}{N_2}\right)}$$

where a is chosen to be an amount slightly less than the observed statistic, d – say $0.9999d$.

2. Let a_k equal the cumulative number of observations from zones 1 through k. For each $k = 1, ..., K - 1$, form the coordinate pairs (i, j), where $i = a_k$ and j takes on all values in the range

$$\max(0, i - N_1) \le j < \text{lower}_a(i)$$

and

$$\text{upper}_a(x) < j \le \min(N_2, i)$$

3. Order all of the pairs into a set $\{B\} = \{i_s, j_s\}$, $s = 1, 2, ..., q$, where these are ordered in terms of increasing i, and increasing j for each i.

4. Find $b_1 = \begin{pmatrix} i_1 \\ j_1 \end{pmatrix}$; $b_s = \begin{pmatrix} i_s \\ j_s \end{pmatrix} - \sum_{r=1}^{s-1} \begin{pmatrix} i_s - i_r \\ j_s - j_r \end{pmatrix}$; $s = 2,...,q$

5. The p-value is then equal to

$$\Pr\left(\chi^2_{max} \geq d\right) = 1 - \left(\frac{N}{N_2}\right)^{-1} \sum_{s=1}^{q} \left(\frac{N - i_s}{N_2 - j_s}\right) b_s$$

6.11 Other Issues with Local Statistics

Some of the questions that arise in the use of local statistic include:

(a) How do we handle the issue of multiple testing? While there may be some cases where there is just a single focal location of interest, it will just as often be the case that we will want to carry out local tests in many locations. We may want to, for example, scan the map, looking for significant local statistics, and of course the multiple testing must be accounted for. These questions will be investigated further in Chapter 7.
How do we define the weights – that is, how do we define the neighborhood, when we don't know which definition will give us the most significant results?

(b) How can we account for global spatial autocorrelation in the data? Suppose we have values ordered in terms of distance around a focal location. Our tests typically assume that observations are independent. But we know that Tobler's first law of geography (Tobler, 1970) says that data are often spatially autocorrelated. Even when there is no raised incidence around a location, data may appear to cluster for no other reason than they exhibit global spatial dependence – high values tend to be near other high values, whether or not there is actually a predisposition for a particular place on the map to have high values. We will address this question both here and in Chapter 8 (where we discuss spatial models).

(c) One of the issues that arises in the use of many local statistics is how to choose the weights. This is equivalent to asking on what spatial scale the raised risk around a site occurs. One approach is to look at multiple neighborhood sizes, and, in the process, account for the multiple testing that occurs. We in fact did this with the cumulative and maximum chi-square tests – we tried different definitions of "near" and "far," took the one that gave us the best results, and then adjusted for the multiple testing through knowledge of the distribution of the test statistic.

In what follows, we will see how the Getis–Ord statistic, when used with binary weights, can be adjusted to account for multiple definitions of scale.

6.11.1 Weights and Multiple Definitions of Scale

How do we choose weights for local statistics? How do we define an appropriately sized neighborhood around the focal location? Weights 0–1 based upon binary adjacency may or may not be the most relevant definition – clusters may exist on smaller or larger scales.

Suppose we have a series of zones around our location, ordered from 1 to n, in terms of increasing distance. We compare Zone 1 (the closest zone) with all other zones. We also compare the first two zones with Zones 3 through n; the first three zones with Zones 4 through n, etc. We could carry out a Getis–Ord test, or, equivalently, a two-sample t-test for each of these neighborhood definitions. From all of these t-statistics, it would be tempting to choose the t-statistic that has the highest absolute value, use the associated weights as the definition of the neighborhood, and compare the statistic with the usual critical value to assess statistical significance. But we need to account for multiple testing. One option would be to use a Bonferroni adjustment – if we carried out s tests, we could use α/s instead of α in determining the critical value. This, however, would be overly conservative – it would raise the critical value more than necessary, and make it too difficult to reject the null hypothesis. Thus the power of the test – the ability to reject false null hypotheses – would be compromised. This is all because the Bonferroni adjustment assumes that the multiple tests are all independent of one another. But clearly the test that compares say Zones 1 and 2 with all others gives a result that is correlated with the result of the test that compares Zones 1 through 3 with all others.

So, we need another approach – one that does in fact account for the multiple testing, but one that does not *over*correct – we are not carrying out s independent tests. We are carrying out r effectively independent tests, where r is unknown, but is somewhere between 1 and s. (Multiple testing is also discussed further in Chapter 7.)

There is a relevant literature on time series (see, e.g., James et al., 1987), specifically where interest is in detecting whether there is a "changepoint", where the mean before the change is different from the mean after the change. One statistic used to evaluate this hypothesis is the maximum of a series of two-sample t-tests, where a set of possible changepoints is examined. We can draw upon this literature that uses aspatial data, and apply it to our spatial problem that seeks to determine whether there is, in effect, a spatial changepoint, dividing zones up into sets that include (a) near locations of interest, and (b) far locations. This is equivalent to examining a range of choices for the definition of the weights. There are several ways to evaluate the statistical significance of the maximum t-statistic.

Our goal here is to find the statistical significance of the maximum Getis–Ord statistic,

$$\max_{k=1,\ldots,n-1} G_k^* \tag{6.29}$$

where there are n regions and here G_k^* is defined around a prespecified location where the weights are equal to one for ordered regions $1, \ldots, k$ and zero for all other regions. This is equivalent to finding the significance of the maximum t-statistic among all possible difference of means tests where we compare the "near" regions with the "far" regions. A discussion similar to the following may be found in Rogerson (2011).

Following the literature aimed at following changepoints in temporal sequences, let $x_1, \ldots, x_n$ be an ordered series of observations; the two-sample difference of means test comparing the first k observations with the remaining observations can be written as

$$G_k = \frac{S_k - \dfrac{kS_n}{n}}{s\left\{\dfrac{k(n-k)}{n}\right\}^{1/2}} \tag{6.30}$$

where k is the number of observations before the changepoint, s is the sample standard deviation of the x's, and S_k is the sum of the first k observations. James et al. (1987) find the distribution of $T_n = \max(G_k)$ and its associated p-value as

$$\Pr\{T_n = \max G_k > b\} = \frac{e^{-\frac{b^2}{2}}\left\{4 + \left(b^2 - 1\right)\ln r\right\}}{4b\sqrt{\dfrac{\pi}{2}}} \tag{6.31}$$

where b is the observed statistic, and

$$r = \frac{t_1\left(1 - t_0\right)}{t_0\left(1 - t_1\right)} \tag{6.32}$$

and $t_0 = 1/n$ and $t_1 = (n - 1)/n$.

Suppose, for example, that we have $n = 100$ observations, and we find a maximum t-statistic of $T_n = 2.95$. If we were only carrying out one t-test this would of course be significant, far exceeding the usual critical value of, say, 1.96. However, we are now accounting for performing 99 tests. In this example we have $r = 99^2$, and $\Pr(T_n > 2.95) = 0.065$. If we were using a Type I error of 0.05, we would (barely) fail to reject the null hypothesis of no changepoint.

One advantage of their formulation is that it is possible to place an upper bound on the number of tests (or equivalently in our case, the number of spatial scales) that are examined. Suppose that we are in a geographical context, and we have 50 zones surrounding a location of interest. We want to know if incidence is raised in the vicinity of this location. We might not necessarily want to try 49 different choices for defining the neighborhood around the location. We might instead want to bound the maximum cluster size so that we are not carrying out so many tests. Choosing an upper limit for the number of tests has the advantage of not robbing the test (whether it is based on a Bonferroni adjustment, or, in this case, the maximum t-statistic) of power. On the other hand, we don't want to constrain our set of neighborhood definitions overly. If we examine only a few choices, we run the risk of not finding larger clusters when they exist.

In Equation 6.31, t_0 and t_1 are the lower and upper limits on the neighborhood sizes that are examined – the expressions given there are the default choices that correspond to testing all possible neighborhood changepoints. We saw that a p-value of 0.065 results when $T_{obs} = 2.95$ and all possible changepoints are examined. In this example there are 100 regions, and we looked at 99 t-tests. Now suppose that we look at only the maximum among the first 40 zones, for our definition of "near" – then $t_1 = 39/100$ instead of $99/100$. Now the p-value is 0.032, a lot less than 0.065 (and, in fact, in this example restriction of the search space causes the reduction in p to take us from accepting the null hypothesis to rejecting it, assuming $\alpha = 0.05$). Viewed from the perspective of critical values, for the full search over the changepoints ranging from 1 to 99, the critical value for $\alpha = 0.05$ is $T_{crit} = 3.05$. When searching only over the first 40 zones for a change, with $\alpha = 0.05$, the critical value is $T_{crit} = 2.77$, making it easier to reject the null hypothesis.

Now consider a comparison between the approach just described and an approach that uses a Bonferroni adjustment. A simulation was carried out and repeated 10,000 times, with a false null hypothesis. For each simulation, $n = 50$ numbers were chosen; these came from a standard normal distribution with mean equal to 0 and variance equal to 1. A random integer between 1 and 49 was then chosen from a uniform distribution; this was the simulated changepoint. Observation numbers after this random integer had one added to their values,. Thus the latter part of the sequence has a higher mean (1) than the early part of the sequence, which has a mean of 0.

Then we test m different possible changepoints. For $m = 1$, we test one possible changepoint – since we don't know where the changepoint is, we use the middle of the series as an estimate, and compare the first 25 observations with the last 25. With $m = 4$, we test possible changepoints after observations 10, 20, 30, and 40. That is, since we don't know the changepoint we spread out the potential changepoints that we do test. This minimizes the correlation between the m tests.

For $m = 1$, a single t-test carried out by splitting the observations into two equal series would find the change 46% of the time. The location of the changepoint may or may not be accurate, but the null hypothesis of a constant mean throughout the series would be rejected almost half of the time. The ability to find the changepoint is non-monotonic in m. That is, the power associated with using a Bonferroni adjustment will increase with m up until a point ($m = 5$ in this example, where the statistical power is 56%, meaning that the null hypothesis is rejected slightly more than half of the time), but then decline. If $m = 6$, one uses $\alpha/6$ to find the critical value, and the resulting power is not quite as high as it is when m is slightly lower. If we used $m = 49$ and tested all possible changepoints, the results would be far too conservative in the sense that the probability of a Type I error would be substantially less than the desired value. For series with n observations, a value of about $0.1n$ seems to be a reasonable choice for m, if a Bonferroni adjustment is to be used. But a better option is to use the null distribution associated with the maximum of the t-statistics, since this accounts for the correlation among the tests. In this example, 59% of the time, the maximum t-statistic exceeds its critical value – better than the power associated with any of the Bonferroni adjustments (but notice – not *that* much better!).

Illustration: Leukemia in central New York State

Data on leukemia are available for a region comprising 280 census tracts, stretching from Binghamton, NY north to Syracuse, NY. The study area has a population of just over a million, and during the period 1978–82, there were 592 cases of leukemia. There was concern over the possible contamination of wells, in particular by the contaminant trichloroethylene (TCE).

(Continued)

There were 11 potential industrial sites of interest and questions centered upon whether the observed number of leukemia cases near any of these industrial sites significantly exceeded what was expected, where the expected numbers are derived by assuming that there was no spatial pattern in the cases relative to population.

Several of the industrial sites are clustered in a small area in and near Binghamton. This points to a difficulty with this type of study. Suppose that one of these industries was responsible for dumping TCEs into the environment, and the risk of leukemia was indeed raised in the vicinity of the site. Because there are other sites nearby, it would be difficult to attribute the problem to one specific industrial site and to rule out nearby sites.

The results are shown in Table 6.4. The first row of the table is Monarch Chemicals, in the Binghamton area. The maximum t-statistic is 4.19 – this is found by comparing the closest $k = 21$ census tracts with the other 259 tracts. If we just look at the first (closest) 50 census tracts as potential definitions of "near" Monarch chemicals, the p-value associated with the maximum t-statistic is 0.001. If we consider the first 150 tracts as potential spatial breakpoints (dividing the area into near and far), the p-value rises only slightly, to 0.002. With these results in hand, we can go back and look at the data. What did we observe in those 21 census tracts? There were 85 cases here – and, importantly, only 48 cases were expected. This difference is unlikely to have arisen by chance. It is more difficult to attribute the problem to Monarch Chemicals, but we can say that the number of cases observed here is significantly more than what would be expected, based upon the population of the area.

Table 6.4   Results for the central New York leukaemia study

Site	max G_i^*	k	$m_1 = \text{max } k$ 50 (p-value)	$m_1 = \text{max } k$ 150 (p-value)	Observed	Expected
1	4.19	21	0.001	0.002	85	48
2	3.38	36	0.018	0.027	122	89
2	3.28	4	0.024	0.037	19	8
2	3.19	108	–	0.047	303	255
3	3.82	44	0.004	0.007	141	96
4	3.86	45	0.004	0.005	144	98
5	2.92	4	0.066	0.097	27	14
6	2.96	22	0.058	0.086	74	49
7	3.22	11	0.029	0.044	50	30
8	3.11	50	0.039	0.059	158	122
8	3.34	147	–	0.031	389	335
9	3.17	42	0.033	0.050	138	104
9	3.30	145	–	0.034	385	331
10	2.95	99	–	0.089	277	230
11	2.43	22	0.203	0.285	63	43

Source: Rogerson (2011)

Some of the locations have p-values less than 0.05; others are greater than 0.05. The p-value is always lower when the scan of spatial scales is restricted to the first 50 zones, versus the first 150 zones. Some of the rows have more than one entry, reflecting secondary maxima in a plot of the difference of means statistic T_k vs. k. Thus the maximum for Location 9 occurred with a neighborhood definition of 145 census tracts. There is a secondary maximum at a distance of 42 tracts. It is possible that some of the significance that occurs at 145 census tracts is coming from some raised incidence that is occurring quite far away, and that raised incidence itself might be "caused" by some other location.

Locations 5 and 7 have smaller clusters. For Location 7, the maximum t-statistic is 3.22. There were 50 cases observed, while only 30 were expected. For Location 5, 2.92 is the maximum t-statistic, and this is achieved when the closest four census tracts are compared with all others. It is barely not significant using 50 tests; it likely would have had a p-value of less than 0.05 if only say the first 10 tracts were tested. The "best" neighborhood definition occurs by aggregating the first four zones. Fourteen cases were expected, and almost twice that many were observed.

6.11.2 Global Spatial Autocorrelation

One of the issues that lurks in the type of testing described above is the possibility of global spatial autocorrelation. The discussion below loosely follows that found in Rogerson (2015b).

Values at one location are often related to nearby values. This can be seen via simulation. Suppose we simulate 50 observations with mean 0 and variance 1, and we calculate the maximum Getis–Ord statistic (which is the same as the maximum t-statistic). If we use the critical values of Yao and Davis (1986), the critical value using $\alpha = 0.05$, not correcting for autocorrelation, is 2.71. If the data are autocorrelated, this is no longer the appropriate critical value. In generating our 50 values, we use

$$x_t = \rho x_{t-1} + \varepsilon_t \tag{6.33}$$

where ρ is a measure of autocorrelation. In addition, we choose x_1 from a standard normal distribution to start the process, and ε is chosen from a normal distribution with mean 0 and variance $1 - \rho^2$. This ensures that the x's will have a distribution with mean 0 and variance 1. When ρ is not equal to 0, an observation is related to, and depends upon, the previous observation. Again, the consequence of assuming independent observations when in fact they are autocorrelated is that too many null hypotheses will be rejected – that is, there will be too many Type I errors. One solution is to use a corrected critical value. Table 6.5, taken from Rogerson (2015b), reports on the results of the simulation. We see that we should for example use a critical value of 3.85 when $\rho = 0.5$.

Table 6.5 Effects of spatial autocorrelation on the Getis–Ord statistic

	Single value of k		Maximum Getis–Ord statistic, across k values	
ρ	Percentage of null hypotheses rejected	Corrected critical value (Type I error probability = 0.05)	Percentage of null hypotheses rejected	Corrected critical value (Type I error probability = 0.05)
0.0	0.050	1.65	0.046	2.71
0.1	0.066	1.80	0.072	2.86
0.2	0.089	1.95	0.108	3.05
0.3	0.101	2.09	0.152	3.30
0.4	0.121	2.28	0.209	3.54
0.5	0.139	2.45	0.276	3.85
0.6	0.166	2.66	0.340	4.19
0.7	0.188	2.95	0.402	4.65
0.8	0.272	3.24	0.463	5.17
0.9	0.253	3.64	0.511	5.83

Note: The second column gives the fraction of (true) null hypotheses rejected, when k is the changepoint using the maximum G-statistic with the critical value of 2.71 that assumes independent observations

Source: Rogerson (2015). Reprinted with permission of John Wiley and Sons.

Following the paper by Henderson (1986), the methods therein were used to estimate and correct for autocorrelation. For the 11 sites, ρ ranges from 0.05 to 0.26. Columns 3 and 4 of Table 6.6 compare z-scores – Column 3 is calculated in the usual way – not adjusting for autocorrelation. Column 4 is a z-score adjusted for autocorrelation, and we see that these values generally have lower absolute values than the z-scores in Column 3. This is consistent with the idea that once autocorrelation has been accounted for, the statistical significance of the results is generally lower. Henderson also provides a means for estimating our confidence in the size of the neighborhood that has been estimated. Location 5 has a value of 0.800; Location 10 has a value of 0.052; these are the probabilities that the estimated maximum is the true maximum.

One can also go back and look at the z-scores associated with the neighborhood definitions. The values in Column 6 represent the z-scores inside of the cluster (i.e., the "close" neighborhood); the values in Column 7 are the z-scores outside of the cluster. We can see that most of the values in Column 6 are positive (i.e., they are above average) and most of the values in Column 7 are negative (i.e., they are below average). For further details, see Rogerson (2015b).

These methods are designed to test the null hypothesis of random values around a location of interest. They allow such testing, albeit accompanied by a host of questions involving weights, global spatial autocorrelation, multiple testing, etc. But in a more general sense, and like so many other exercises in statistical and spatial statistical analysis,

Table 6.6 Results for 11 focal locations in central New York

Location (1)	Autocorrelation parameter (2)	Chernoff–Zacks (3)	Henderson's statistic (equation (17)) (4)	Optimal k (obs. no.) (5)	Mean "close" (6)	Mean "far" (7)	Posterior probability of change at k (8)
1. Monarch	0.18	1.00	0.24	21	1.12	−0.02	0.235
2. IBM Endicott	0.11	2.64	2.20	4	2.05	0.41	0.290
3. Singer	0.17	2.97	2.13	44	0.69	−0.72	0.295
4. Nesco	0.11	2.85	2.29	45	0.69	−0.99	0.354
5. GE Auburn	0.24	−2.00	−1.76	4	1.83	−0.41	0.800
6. Solvent Savers	0.13	−0.34	−0.58	22	0.78	−0.32	0.106
7. Smith Corona	0.23	−1.98	−1.85	11	1.21	−0.30	0.154
8. Victory Plaza	0.17	3.46	2.33	1	2.78	0.48	0.252
9. Hadco	0.26	3.39	2.55	1	−0.84	0.56	0.068
10. Morse Chain	0.05	1.11	0.96	43	0.25	−0.67	0.052
11. Groton	0.17	−0.49	−0.52	22	0.65	−0.37	0.148

Note: The statistics in Columns 3 and 4 are standard normal variates (z-scores) under the null hypothesis of no changepoint

Source: Rogerson (2015). Reprinted with permission of John Wiley and Sons.

they also help to organize one's thinking and provide a basis for the next set of questions to ask. We of course wish to take the specific results seriously, but, in addition, we want to use the analysis to put into context the questions we are asking, the limitations of the methods and of the data, and to formulate additional investigations.

Exercise

For the following data on the observed and expected number of cases for a set of six regions (where the regions appear in order of increasing distance from the focal point of interest), carry out (a) Stone's test, (b) the score test, (c) the score test using modified weights (i.e., dividing the weight w_{ij} by the square root of the expectation of j), and (d) the local Getis–Ord G statistic, using z-scores formed by $(O_i - E_i)/\sqrt{E_i}$. For this question, use the set of weights given below, and then repeat, using weights of 1 for the first two regions, and weights of 0 for the other regions.

Observed	Expected	Weight
15	10	1
14	10	0.5
12	10	0.4
10	13	0.3
10	12	0.2
4	10	0.1

Solutions

(a) Stone's test: Statistic is equal to 1.5.
(b) Score statistic: 5.9. Variance = 5.21; z = 2.59; reject null hypothesis
(c) Local spatial chi-square: 15.41; Variance = 25.16; z = 2.60; reject null hypothesis
(d) z-scores {1.58 1.26 0.632 –0.832 –0.577 –1.90}; mean = 0.0286; standard deviation = 1.35; G = (1.911 – 2.5 × .0287) / {1.35 [(6 × 1.55 – 6.25)/5]$^{1/2}$} = 1.744

Further reading

Classic early references to local statistics are the papers by Anselin (1995), Getis and Ord (1992), and Ord and Getis (1995). Waller and Gotway (2004) and Dale and Fortin (2014) also have good expositions of these topics.

Online resources

Visit **https://study.sagepub.com/rogersonspatialstatistics** for resources which supplement the material presented in this chapter. Students can find datasets, additional exercises and sample code files, which can be used to practice the techniques covered in this chapter, as well as a chapter introduction video and video demonstration of executing Local Moran's I and getting the Getis–Ord Gi* statistic using ArcGIS. The topic-software chart also describes the software packages that can be used to implement key techniques.

Lecturers and instructors can find exercises and conceptual prompts to facilitate classroom discussion and practice.

7

Tests for the Detection of Clustering: Scan Tests

Chapter overview

In this chapter, you will learn about:

- Issues associated with multiple testing
- An aspatial scan test
- Kullorff's spatial scan statistic
- Gaussian scan statistic
- Rectangular scan statistic

7.1 Introduction: Multiple Testing

It will often be of interest to derive local statistics for many, if not all, of the regions comprising a study area. However, when carrying out multiple local tests, the multiple testing needs to be accounted for.

Suppose you carried out local tests for all m regions as possible focal regions. A natural inclination is to examine the list of local statistics for all regions, and focus on the one that has the largest value (and hence the smallest p-value). To evaluate that, though, we have to account for the fact that we "cherry-picked" our results – we chose the one that was most significant. If for example there were 100 two-tailed z-tests, we would compare the maximum observed z-statistic with a critical value of 3.29 that is found via a Bonferroni adjustment by using $\alpha = 0.05/100 = 0.0005$, instead of the "usual" critical value of 1.96 that is associated with $\alpha = 0.05$. This of course makes rejection of the null hypothesis relatively more difficult. This means that it will also be harder to reject null hypotheses when they are false – the statistical power could be very low. One assumption made in employing the Bonferroni adjustment is that the m statistical tests are *independent*. If the tests are correlated with one another (and as we will see, with local statistics for adjacent regions using much of the same information, they often are), then the Bonferroni adjustment is too conservative – that particular adjustment results in a critical value that is too high, and this makes it too difficult to reject null hypotheses. If we are not doing m independent tests, and there is overlap between tests, there are say m' effectively independent tests, where m' is some number between 1 and m. How to find m' is not necessarily clear or straightforward in all cases, but we will look at some solutions to this question. Developing and using methods for addressing the issue of multiple testing is an active area of current research (see, e.g., the literature on the false discovery rate (Benjamini and Hochberg 1995; Benjamini 2010); although beyond the scope here, this is an important new area of research).

There are also more difficult and deeper issues associated with multiple testing as well. Suppose you have a map of the United States, and you examine cancer rates for all 3,143 counties. Comparing observations with expectations (i.e., expectations for the number of cases, if there is no raised incidence), let's say one finds that there are 150 counties with z-scores over 1.645, but all are under the critical value of 4.16, which is associated with $\alpha/3,143$ and the simultaneous examination of 3,143 tests. The national analyst concludes that there is not enough evidence to be concerned – one would after all expect about 1/20th of the counties to exceed the critical value of $z = 1.645$ by chance alone. Now suppose that you are a county health commissioner in any one of those 150 counties that have z-scores greater than 1.645. As health commissioner, you are only concerned with one test – namely, the one associated with your own county. You of course find that the rate *is* statistically significant – after all, the z-score is greater than 1.645, and it would not be expected very often by chance alone, if there were no true elevation in the rate. How does one reconcile the differing conclusions reached by the county health commissioner and the national analyst? There is not an easy answer to that question.

In this chapter we introduce several approaches to questions involving multiple testing – focusing in particular on scan tests that can essentially be viewed as "moving windows," where tests are carried out as the window moves across the study area.

7.2 Basic Aspatial Scan Test – Maximum of Aspatial z-scores (M-test)

The local quadrat scan test (also known as the M-test; see Fuchs and Kennet, 1980) compares the maximum z-score that is observed with the critical value based upon a Bonferroni adjustment.

Suppose we randomly scatter 80 points across 25 cells or regions. We would expect an average of 80/25 = 3.2 per cell. For each cell, we can compute a z-score, equal to the observed count in the cell, minus 3.2, with that result being divided by the square root of 3.2. How can we evaluate the maximum of the 25 z-scores? Note in this example that there are no spatial effects. Alternatively stated, all of the local statistics are based upon only the information from their own cells – weights associated with other cells are all equal to zero. We wish to scan the map, a cell at a time, and search for cells that have more points than could be expected by chance alone.

Typically we are interested in a one-tailed test of the null hypothesis that the points are randomly distributed across cells, against the one-sided alternative hypothesis that there is at least one cell where there are *more* points in a cell than we expect. We can search for cells with more points than we expect by using this local quadrat test. And, as an aside, although the usual alternative hypothesis is set up to look for cells with more events than expected (i.e., hotspots or outliers), it is certainly possible to look for "cool" spots, where the number of observations in a cell is *less* than what one would expect by chance alone.

Returning to our example, the observed z-score for each cell would be compared with a critical value associated with the standard normal deviate that leaves an area of 0.05/25 in the right tail of the distribution. This is equal to a critical value of $\Phi^{-1}(1 - 0.05/25) = -\Phi^{-1}(0.05/25) = 2.878$, where the notation $\Phi^{-1}(x)$ refers to the value of z associated with an area to the left of x that is under the standard normal distribution. A quick calculation shows that nine points would be required in the cell; that is, the minimum integer value that would result in a z-score greater than 2.878 if the expected number of points in the cell were 3.2 (Rogerson and Yamada, 2009).

The z-score in each quadrat can be derived from either an unconditional perspective, or a conditional one (recall the discussion in Section 6.2). This does not change the method, but it does have a small impact on the results. In particular, when the unconditional model is used (e.g., viewing each cell or quadrat count as coming from a Poisson or binomial distribution, with no constraint on the total count), the z-distribution can be viewed as the common approximation to the Poisson or binomial. Fuchs and Kennet use the binomial distribution, defining the value in cell i as

$$z_i = \frac{x_i - Np_i}{Np_i(1 - p_i)} \tag{7.1}$$

in the conditional case where the total number of points across cells is fixed. Here x_i is the number of points in cell i, p_i is the probability that a point falls in cell i, and N is the total number of points. For their conditional case, the Bonferroni-derived critical value is a conservative upper bound on the actual critical value – there are correlations between the individual quadrat tests (if there are more counts in one cell, there are likely fewer in others) and Fuchs and Kennet use these correlations to also derive a lower bound for the critical value.

What about the power of this test – and, in particular, how does the ability of the M-test to find deviations from expectations differ from the global (aspatial) chi-square test? When there is just a single outlying cell with observations exceeding expectations, the M-test will have relatively better power in detecting it. If there are just slight excesses in each of many regions, then the global chi-square test will do better than the M-test. Fuchs and Kennet suggest that when the number of outliers is less than about 5%–10% of the number of cells, the M-test will be relatively better than the global test.

Although the M-test is aspatial, there are several alternative ways in which geography has been injected into the search for regions of raised incidence on maps. In this chapter we focus on two such methods – Kulldorff's spatial scan statistic and a Gaussian scan statistic (along with a brief look at a rectangular scan statistic). Before doing so, however, early contributions to the field of scan statistics for geographical applications should be acknowledged. Openshaw et al.'s Geographical Analysis Machine (1987) placed circles of various sizes around many grid points that were overlaid onto the study area. Multiple testing was accounted for in a crude way, by lowering the usual p-value to an arbitrary amount. Fotheringham and Zhan (1996) modified this approach by reducing the number of circular regions examined, choosing arbitrary centers and radii for them. Besag and Newell (1991) examined the sizes of circular regions necessary to accumulate a given number of cases, and if these were sufficiently small, they were declared significant. Turnbull et al. (1990) created circular regions of equal population around multiple locations; the most significant cluster was the one with the most cases. Significance was assessed via simulation. Rushton and Lolonis (1996), in an application to the spatial pattern of birth defects, also formed circular regions, all of a given radius, around grid points that were a specified distance apart – they compared the number of cases observed within the circles with those derived from a simulation of the null hypothesis that the risk of birth defects was spatially uniform. Their output was in the form of a map of p-values derived from the comparison.

7.3 Kulldorff's Spatial Scan Statistic

Kulldorff's spatial scan statistic is by far the most frequently used method to find spatial clusters. Its popularity has been boosted through the availability of the freely downloadable software, SATSCAN. The statistic is based upon a likelihood ratio. Before describing the specifics of the test, we first introduce the ideas and methods associated with likelihood ratio tests.

7.3.1 Likelihood Ratio Tests

The likelihood ratio is formed by dividing the likelihood of the data under the null hypothesis by the likelihood of the data under an alternative hypothesis:

$$\lambda = \frac{\mathcal{L}_0}{\mathcal{L}_1} \tag{7.2}$$

This ratio will always lie on the interval between 0 and 1. For large sample sizes, twice the negative log of the likelihood ratio (using natural logs) has a chi-square distribution, with degrees of freedom equal to the difference in the number of parameters estimated when finding $\mathcal{L}_0$ and $\mathcal{L}_1$. The likelihood ratio statistic, LR, can therefore be expressed as twice the difference in the log likelihoods

$$LR = -2\ln \lambda = -2\ln \frac{\mathcal{L}_0}{\mathcal{L}_1} = -2\{\ln(\mathcal{L}_0) - \ln \mathcal{L}_1\} = 2\{\ln \mathcal{L}_1 - \ln \mathcal{L}_0\} \tag{7.3}$$

Example

Suppose there is interest in determining whether there is raised risk of disease in the census tract containing a hazardous waste site, relative to the adjacent tracts. In the tract containing the hazardous site (Region 1), three cases of disease are observed among 20 people; in the surrounding tracts (Region 2) there are four cases of disease among 60 people. Use the likelihood ratio test to test the null hypothesis that the risk of disease is spatially uniform.

The data may be expressed in the 2 x 2 table (Table 7.1).

Table 7.1   Cases of disease in two regions

	Disease	No disease	Total
Region 1	3	17	20
Region 2	4	56	60
Total	7	73	80

Let the notation for the frequencies that are observed and expected (under the null hypothesis of no effect of region on the disease rate) be as given in Table 7.2.

Table 7.2   Observed and expected cases in the two regions

	Observed	
	Disease	No disease
Region 1	o_1	o_3
Region 2	o_2	o_4
	Expected	
	Disease	No disease
Region 1	e_1	e_3
Region 2	e_2	e_4

Under the null hypothesis, all individuals would have an equal likelihood of disease, and this would be equal to $p = 7/80$, since there are a total of m 80 people and 7 with disease. The likelihood of the data is given by the binomial distribution, and is proportional to $p^7(1-p)^{73}$ (the combinatorial $\binom{80}{7}$ that precedes the binomial expression is unnecessary here, since it is a constant). The alternative is that there are separate risks of disease in the two regions – p_1 = 3/20 in the tract containing the hazard, and p_2 = 4/60 in the other tracts. In this case the likelihood of the data is proportional to $p_1^3 (1-p_1)^{17} p_2^4 (1-p_2)^{56}$. The likelihood ratio is

$$\lambda = \frac{\mathcal{L}_0}{\mathcal{L}_1} = \frac{p^7 (1-p)^{73}}{p_1^3 (1-p_1)^{17} p_2^4 (1-p_2)^{56}} \tag{7.4}$$

and the likelihood ratio statistic, which is equal to twice the negative log-likelihood ratio is

$$LR = -2\ln\lambda = 2\{\ln\mathcal{L}_1 - \ln\mathcal{L}_0\}$$
$$= 2\{3\ln p_1 + 17\ln(1-p_1) + 4\ln p_2 + 56\ln(1-p_2) - 7\ln p - 73\ln(1-p)\} \tag{7.5}$$

The reader may wish to verify that this implies that LR = 1.174. This is to be compared with a critical value from a chi-square table having one degree of freedom (there are two parameters associated with the alternative hypothesis, and one parameter associated with the null hypothesis; the difference in the number of parameters is equal to the degrees of freedom associated with the likelihood ratio test). With α = 0.05, the critical value of a chi-square variable with one degree of freedom is equal to 3.84. Since LR < 3.84, we do not have enough information to reject the null hypothesis of a uniform disease rate across the entire study area.

The likelihood ratio test yields a result that is similar to the chi-square goodness-of fit test. As we saw in Chapter 6, the chi-square goodness-of-fit statistic compares the observed and expected number of cases in each region. The expected number of cases (e) in each place (under the null hypothesis) is equal to the overall individual probability (7/80), multiplied by the number of people in that place. Thus e_1 = (7/80)(20) = 1.75 and e_2 = (7/80)(60) = 5.25. Note that the total number of cases expected (5.25 + 1.75) is equal to the total number of observed cases:

$$\chi^2 = \sum_{i=1}^4 \frac{(o_i - e_i)^2}{e_i} = \frac{(3-1.75)^2}{1.75} + \frac{(4-5.25)^2}{5.25} + \frac{(17-18.25)^2}{18.25} + \frac{(56-54.75)^2}{54.75} = 1.305 \tag{7.6}$$

The likelihood ratio is

$$\lambda = \frac{\mathcal{L}_0}{\mathcal{L}_1} = \frac{(e_1/m)^{o_1} (e_2/m)^{o_2} (e_3/m)^{o_3} (e_4/m)^{o_4}}{(o_1/m)^{o_1} (o_2/m)^{o_2} (o_3/m)^{o_3} (o_4/m)^{o_4}} = \frac{e_1^{o_1} e_2^{o_2} e_3^{o_3} e_4^{o_4}}{o_1^{o_1} o_2^{o_2} o_3^{o_3} o_4^{o_4}} \tag{7.7}$$

The likelihood ratio statistic is

$$LR = -2\ln \lambda = -2\sum_{i=1}^{4} o_i \ln\left(\frac{e_i}{o_i}\right) = 2\sum_{i=1}^{4} o_i \ln\left(\frac{o_i}{e_i}\right) \qquad (7.8)$$

For the question above, $LR = 1.174$; this value is identical to that derived above. This likelihood ratio statistic is known as the G-statistic, and its general form when there are k categories is

$$LR = G = -2\ln \lambda = 2\sum_{i=1}^{k} o_i \ln\left(\frac{o_i}{e_i}\right) \qquad (7.9)$$

Under the null hypothesis that observations do not differ significantly from expectations, G has a chi-square distribution with $k - 1$ degrees of freedom. Asymptotically (i.e., for large sample sizes), G and the chi-square test give identical results. This test is commonly used instead of the chi-square goodness-of-fit test in fields such as biometry and biology; see, for example, Sokal and Rohlf (1981) and McDonald (2014).

The likelihood ratio test just described is for a single test of the null hypothesis of no raised incidence around one specific location– for example, it may be used if we want to know whether the probability of being a disease case for locations within 2 miles of a point of interest (e.g., a hazardous waste site) is higher than the probability of being a case for locations greater than 2 miles from that point.

We next give an example to show how these notions of likelihood may be extended and used in geographic cluster detection. The likelihood ratio test described above will be carried out many times – for different spatial definitions of "near" and "far" for specific locations on the map. This may be conceptualized in the form of an expanding circle around a site, where the area within the circle is defined as "near" the site. This scenario was discussed in Chapter 6, where the assessment of multiple spatial scales for local statistics was discussed. This process is then repeated for all locations on the map. Multiple testing thus crops up in two ways – for different spatial scales around a particular site, and also in the form of scanning multiple locations across the map. Kulldorff's spatial scan statistic (see, e.g., Kulldorff, 1997) is the maximum among all of the likelihood ratios associated with these tests. We of course want to account for the multiple testing in an assessment of statistical significance, and this is done by simulating the null hypothesis.

The following discussion follows closely that in Kulldorff and Nagarwalla (1995). Suppose that we have C cases and $N - C$ controls. We wish to find the geographic region where the likelihood of being a case is high (relative to the likelihood of being a case, outside of that geographic region). Let p be the probability that an observation is a case inside of a defined zone z, and q be the probability that an observation is a case outside of zone z. Then the likelihood is proportional to

$$L = p^{c_z}(1-p)^{n_z - c_z} q^{C - c_z}(1-q)^{(N - n_z) - (C - c_z)} \qquad (7.10)$$

where c_z and n_z are the number of cases and observations inside zone z, respectively. We do not know p and q, but they can be replaced with their maximum likelihood estimators of c_z/n_z and $(C-c_z)/(N-n_z)$, respectively (for more detail on maximum likelihood estimators, see Appendix A). Kulldorff and Nagarwalla seek the zone z that maximizes this likelihood. That is, the most likely cluster is the zone z^* for which L is a maximum. This can be carried out in practice by locating circular zones of increasing sizes around sites within the study area – and then repeating this by scanning the map (i.e., repeating the analysis for other locations on the map).

The likelihood of the data under the null hypothesis, L_0, is proportional to $p^C(1-p)^{N-C}$, where $p = C/N$. Although this could be used together with Equation 7.10 to form a likelihood ratio, since L_0 is constant across all tests, it suffices to maximize L in Equation 7.10.

The observed maximum likelihood is compared with a distribution of simulated likelihoods found by simulating the null hypothesis. That is, each person is assigned the disease with probability $p = C/N$. Then the maximum likelihood across all definitions of zones is found. This is repeated a large number of times (say T) to generate a distribution of maximum likelihoods. The 95th percentile of this distribution can serve as a critical value, for comparison with the observed value of the maximum likelihood. Similarly if there are t^* simulated likelihoods that are higher than the one observed, the p-value associated with the null hypothesis of no clustering is equal to t^*/T.

In addition, in practice one uses the natural log of L (i.e., $\ln L$) instead of L (the solution is unaffected when one maximizes the natural log of the likelihood instead of L).

7.4 A Gaussian Scan Statistic

We now look at a method for spatial cluster detection that is based upon an analogy with the medical imaging literature that concerns brain activation studies. Following specific tasks, there can be increases in blood flow in parts of the brain – parts of the brain that are "activated" and researchers wish to know what parts of the brain are associated with tasks of various kinds. Three-dimensional models of the brain consisting of a large number of three-dimensional, cubic building blocks (termed voxels) can be used to keep track of any changes in blood flow, and the goal is to find significant areas of brain activation – these are three-dimensional spatial clusters of significantly increased blood flow. Worsley (1996) developed methods using principles from the field of integral geometry to evaluate the statistical significance of clusters in d dimensions.

Our general objective here is to utilize this approach in a two-dimensional spatial setting, and to evaluate the significance of the maximum of a set of correlated local statistics. If the local statistics were independent, a simple Bonferroni adjustment would suffice. We'll take regional values and assume they are normally distributed (if not, we can transform them to approximate normality). The next step is to create local statistics that are a special case of the Getis–Ord statistic by using weights that have a profile mimicking the normal distribution (this is known as a set of weights that constitute a Gaussian kernel). To summarize, we have normally distributed regional values (and, in addition, we are going to look at this for applications where the two-dimensional map consists of a set of square regions). We smooth those regional values to create local statistics using weights with a normally distributed profile (i.e., a Gaussian kernel). We then want to evaluate the peak of this smoothed map by asking whether the peak is likely to have occurred by chance alone, if in fact the null hypothesis of no clustering holds.

The methods described here can be thought of in a few different ways. One is to imagine a kernel density map – a smoothed version of a map of points or regional values. Most GIS packages produce nice visualizations like this. Even when there is no spatial clustering or raised incidence, we don't expect these maps to be perfectly smooth, and our goal is to evaluate the statistical significance of the highest peak on such a smoothed map. While visually pleasing, and informative in giving the viewer some idea of spatial variation, there is typically no way to evaluate the statistical significance of the geographic variation, and the importance of the peaks. Worsley's contributions give us an analytical method that can be used to ask how unusual the peaks on such maps are, assuming the null hypothesis of no raised spatial variation (where only stochastic variability would be expected).

This approach may also be viewed as one where we are looking at z-scores for regions – those z-scores could be defined for example as $z = $ (observed – expected) / $\sqrt{\text{expected}}$. Then these z-scores are turned into local statistics by looking at a weighted sum of the z-scores in the vicinity of a region. The weights are based upon a Gaussian kernel. Finally, we are interested in evaluating the statistical significance of the maximum of these local z-scores. In what follows we assume that values are observed for a set of regular, square cells. This would be typical for example of remotely sensed images that are based on a set of values for pixels. We can thus view the method here as one that is a spatial version of the M-test described in Section 7.2 – instead of evaluating the significance of a z-score for each individual region, we will be evaluating z-scores that have been constructed by weighting data from surrounding regions as well.

We will be able to use analytical solutions and approximations to find critical values and p-values.

To find an appropriate critical value, one idea is to use a Bonferroni adjustment – this would be appropriate if there were no smoothing or weighting, and if all cells were independent. This would entail basing the critical value on α/m, where m is the number of cells. In reality, though, in evaluating all of the local statistics, we must recognize that these local statistics are correlated. Two adjacent regions will have similar local statistics, because they are both constructed using a weighted sum of surrounding z-scores. Thus, a better adjustment would be to divide α by something less than m (but more than 1). We are doing more than one test, but we are doing fewer than m independent tests. We are carrying out a number of "effectively independent" tests, and this is less than m.

A summary of the approach is as follows:

1. Start with a grid (lattice) of m cells, filled with standardized values (z) that are assumed to have come from a normal distribution. For each cell i, find a local statistic (y) that is a weighted sum of that cell and surrounding cells j, using weights from the Gaussian kernel:

$$w_{ij} = \left(\sqrt{\pi}\sigma\right)^{-1} e^{-\frac{d_{ij}^2}{2\sigma^2}}$$ (7.11)

where σ is the standard deviation (also termed the bandwidth) that determines the amount of "smoothing" of the map of z's. The distance from i to j is given by d_{ij}, and is measured in units of grid cell length – for example, $d_{ij} = 1$ is the distance from the center of a cell to the center of any of the four cells to the north, east, south, or west. Thus the cells are scaled to have unit length.

2. Form the local statistics $y_i = \sum_{j=1}^{m} w_{ij} z_j$. The resulting local statistics have a standard normal distribution. The Gaussian scan statistic is the maximum of the local statistics, $\max_{i=1,\dots,m} y_i$.

3. Compare the observed Gaussian scan statistic with the critical value y^* found from

$$y^* = \sqrt{-\sqrt{\pi}\ln\left(\frac{3.24\alpha\sigma_t^2}{m}\right)}$$ (7.12)

where m is the number of regions or cells, and $\sigma_t^2 = \sigma^2 + 1.11^2$.
Further details are discussed in Rogerson (2001).

These results come from work on the geometry of random fields. A random field is a function that takes on values at each point in an area. Worsley derived exceedance probabilities – these are likelihoods that a field of values would exceed and rise above a horizontal plane set at a critical value. Furthermore, this was carried out for different dimensional spaces. For two dimensions, he finds that the probability that the maximum local statistic exceeds a value z^* is

$$\Pr\left(Z > z^*\right) = \frac{Az^*\varphi\left(z^*\right)}{4\pi\sigma^2} + \frac{D\varphi\left(z^*\right)}{\sqrt{\pi}\sigma} + \left(1 - \Phi\left(z^*\right)\right)$$ (7.13)

where A is the number of cells, D is the square root of the number of cells, and φ and Φ are, respectively, the probability density and cumulative distribution function associated with the standard normal distribution.

Rogerson (2001) describes the following illustrative experiment. Take a 30 × 30 grid of cells, and put z-scores in each cell. Then create local statistics, or, equivalently, smooth the z-scores, by using a normal (Gaussian) kernel with chosen values of σ = 1, 2, and 3. Here sigma is scaled: σ = 1, for instance, means that most of the area under the kernel is within 2 standard deviations, or 2 grid cells from the focal location. With σ = 2, most of the area would be within 2 standard deviations, or 4 grid cells. Equivalently, most of the weight would be assigned to these cells. There are also edge effects – the z-scores in the corners and on the edges of the grid are not smoothed as much, since they are not surrounded by as many cells as those cells in the interior of the grid. To avoid these effects, only the resulting smoothed z-scores constituting the local statistics in the 22 × 22 interior portion of the grid were used to report results. The 95th percentile of the distribution of the maximum local statistic can serve as a critical value. These percentiles were equal to 3.575, 3.342, and 3.110 for the cases of $\sigma = 1$, 2, and 3, respectively. With 22 × 22 = 484 observations, we don't want to use $z = 1.96$ as a critical value – too many null hypotheses would be rejected. A Bonferroni adjustment for 484 tests with $\alpha = 0.05$ would correspond to a critical value of $z^* = 3.711$. Tests are highly correlated when we do a lot of smoothing – this reduces the effective number of tests, and leads to critical values less than 3.711.

When we set the left-hand side of Equation 7.13 equal to 0.05, and we use $z = 3.711$, we can solve the equation for σ – this will tell us the amount of smoothing implicit in using a discrete grid. The value of σ in this case turns out to be 1.11. This is the additional amount of smoothing that we want to account for because of the discreteness of space. The total amount of smoothing σ_t^2 is therefore equal to the value of σ associated with the Gaussian kernel, plus this discrete grid correction:

$$\sigma_t^2 = \sigma^2 + 1.11^2 \tag{7.14}$$

Next, we can attempt to simplify the equation for the critical value. The terms in the equation are decreasing in magnitude (the reader may wish to verify this by using a value of z, a value for the number of square cells (A), and recognizing that D is equal to the square root of A). Therefore, we can get a good approximation by using two terms instead of all three on the right-hand side of Equation 7.13, and, in fact, a reasonable approximation also results from just using the first term.

To find the number of effectively independent tests we'd like to divide α by something – by something greater than 1 (we're doing more than one test), and something less than 484 (since the tests are not independent). How many effectively independent tests are we carrying out? Let's call that number r; the higher σ is, the more smoothing and the lower r should be. We know the number of cells (m), and we know σ – these values were used in a simulation to estimate a constant of proportionality k in the equation

$$r = \frac{m}{(k\sigma_t)^2} \tag{7.15}$$

Different values of k were tried and a value of $k = 0.9$ consistently worked well – that is, that particular value of k yielded a value of the effective number of independent tests (r) that could then be used in a Bonferroni adjustment to yield a critical value that in turn would reject just the right proportion (namely, α) of tests, when the null hypothesis were true.

Note that the denominator of the expression in Equation 7.15 may be rewritten as

$$k^2\sigma_t^2 = k^2\left(\sigma^2 + 1.11^2\right) = 1 + 0.81\sigma^2 \qquad\qquad (7.16)$$

So, when $\sigma^2 = 0$, there is no smoothing and $(k\sigma)^2 = 1$; in this case $r = m$, and this is equivalent to the Bonferroni adjustment, for m independent tests. To take another example, if $m = 484$ and $\sigma = 1$, then $r = 484 / (1 + 0.81) = 484 / 1.81 = 267.4$.

Finally, there is a way to avoid the use of the standard normal distribution table. Specifically one can use an approximation for the cumulative standard normal distribution (Sasvári and Chen, 1999) together with the fact that we would like to leave α/r in the tail of the standard normal distribution, Rogerson (2001) shows that we arrive at Equation 7.12.

7.4.1 Some Spatial Applications

The Gaussian scan statistic was applied to the study of cerebrovascular disease (stroke) in New York State (Han et al., 2005). Data on stroke were available by zip code, and there was interest in the effect of income on stroke rates. The initial hypothesis was that there was a U-shaped relationship between stroke and income. Thus rates were expected to be high for individuals and households with low income. And although rates do decrease with increasing income, the working hypothesis was that above a certain income level, stroke rates would begin to increase. In part this could be due to lifestyle factors – at levels of income that are sufficiently high, factors such as diet (e.g., eating out a lot) and lack of exercise may begin to be associated with increased risk of stroke. Of course there are many factors that are known to affect stroke risk – education, diet, smoking, etc. This particular analysis focused upon the effects of income alone, and as a byproduct, the spatial pattern of the disease across zip codes in New York State was analyzed. The SPARCS data set was used to look at incidence (not mortality). A regression model was used to look at the relationship between age-adjusted stroke rates and income. The nonlinear effect with income was confirmed, and the residuals from the regression were calculated and smoothed using Gaussian weights. The smoothed residuals were compared with the critical values associated with the Gaussian scan statistic. The geographic areas of statistical significance (for both high and low values) are shown in Figure 7.1. There are two regions with high rates – southern Long Island, and Buffalo. These rates are higher than can be expected, based upon the age distribution and income in these regions.

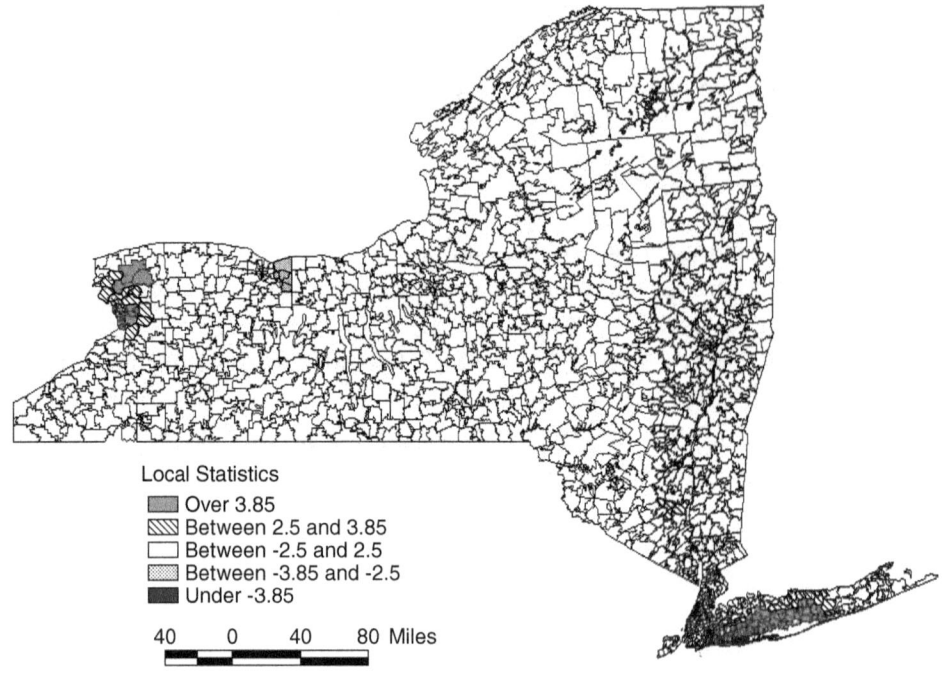

Figure 7.1 Applications of the Gaussian scan statistic

Source: Han et al. (2005). Reprinted under a Creative Commons license 4.0.

Other applications of the Gaussian scan statistic include the spatial analysis of prostate cancer in the United States (Rogerson et al., 2006) and breast cancer (Han and Rogerson, 2003).

7.5 A Simple Gaussian Scan Statistic

Rogerson and Wang (2013) showed how Worsley's scan statistic can be implemented simply for the detection of clustering in two dimensions when the study region consists of a set of square cells.

For the special case when the bandwidth (i.e., the standard deviation of the Gaussian kernel smoother) is equal to 1, the weights in Equation 7.11 simplify to

$$w_{ij} = \frac{1}{\sqrt{\pi}} e^{-\frac{d_{ij}^2}{2}} \qquad (7.17)$$

Now suppose that the Gaussian kernel is centered on cell i. The distances from i to surrounding cells are given in Figure 7.2. These distances are in "grid cell" units; that is, the distance from one cell to an adjacent cell is equal to one. If we let z_j represent the data in cell j (assumed to be from a standard normal distribution), then the local statistic for cell i is a weighted sum of the z's:

$$y_i = \frac{1}{\sqrt{\pi}}\left\{z_i + e^{-1/2}\sum_{j=1}^{4}z_j^{(1)} + e^{-1}\sum_{j=1}^{4}z_j^{(\sqrt{2})} + e^{-2}\sum_{j=1}^{4}z_j^{(2)} + e^{-5/2}\sum_{j=1}^{8}z_j^{(\sqrt{5})} + e^{-4}\sum_{j=1}^{4}z_j^{(2\sqrt{2})} + \dots\right\} \tag{7.18}$$

where the parenthetical superscript notation "(d)" refers to one of the cells that is at such distance from cell i. There are of course more terms in the summation and these correspond to weights associated with more distant cells, but these weights are small enough to be ignored in this case.

$$\begin{bmatrix}
\sqrt{18} & \sqrt{13} & \sqrt{10} & 3 & \sqrt{10} & \sqrt{13} & \sqrt{18} \\
\sqrt{13} & \sqrt{8} & \sqrt{5} & 2 & \sqrt{5} & \sqrt{8} & \sqrt{13} \\
\sqrt{10} & \sqrt{5} & \sqrt{2} & 1 & \sqrt{2} & \sqrt{5} & \sqrt{10} \\
3 & 2 & 1 & 0 & 1 & 2 & 3 \\
\sqrt{10} & \sqrt{5} & \sqrt{2} & 1 & \sqrt{2} & \sqrt{5} & \sqrt{10} \\
\sqrt{13} & \sqrt{8} & \sqrt{5} & 2 & \sqrt{5} & \sqrt{8} & \sqrt{13} \\
\sqrt{18} & \sqrt{13} & \sqrt{10} & 3 & \sqrt{10} & \sqrt{13} & \sqrt{18}
\end{bmatrix}$$

Figure 7.2 Distances from i to surrounding cells

Substituting the values of the constants e and π, we can write the local statistic as

$$y_i = 0.56z_i + 0.34\sum_{j=1}^{4}z_j^{(1)} + 0.21\sum_{j=1}^{4}z_j^{(\sqrt{2})} + 0.08\sum_{j=1}^{4}z_j^{(2)} + 0.05\sum_{j=1}^{8}z_j^{(\sqrt{5})} +$$
$$0.01\sum_{j=1}^{4}z_j^{(2\sqrt{2})} \tag{7.19}$$

In alternative parlance, constructing the local statistics in this way can be described as passing a 5×5 filter over the study area, successively centering the following symmetric weight matrix or kernel over each cell:

$$\begin{bmatrix}
0.01 & 0.05 & 0.08 & 0.05 & 0.01 \\
0.05 & 0.21 & 0.34 & 0.21 & 0.05 \\
0.08 & 0.34 & 0.56 & 0.34 & 0.08 \\
0.05 & 0.21 & 0.34 & 0.21 & 0.05 \\
0.01 & 0.05 & 0.08 & 0.05 & 0.01
\end{bmatrix}$$

The critical value is also found by substituting $\sigma = 1$ into Equation 7.12:

$$y^* = \sqrt{-1.77\ln\left(\frac{7.24\alpha}{m}\right)} \tag{7.20}$$

where m is the number of cells in the study region. This expression may be rearranged to find the p-value associated with the maximum local statistic, y_{max}:

$$p = \left(\frac{m}{7.24}\right) e^{-\frac{y_{max}^2}{1.77}}$$ (7.21)

A similar simplification results from using $\sigma = 0.7$ in the expression for the weights, so that the weights are close to round, single-digit numbers. In this case

$$y_i = 0.8z_i + 0.3\sum_{j=1}^{4} z_j^{(1)} + 0.1\sum_{j=1}^{4} z_j^{(\sqrt{2})}$$ (7.22)

where the weights have been rounded just slightly to maintain simplicity. This can be expressed in the form of the 3×3 filtering matrix:

$$\begin{bmatrix} 0.1 & 0.3 & 0.1 \\ 0.3 & 0.8 & 0.3 \\ 0.1 & 0.3 & 0.1 \end{bmatrix}$$ (7.23)

Expressions for the critical value y^* and the p-value are found by using Equation 7.12 with $\sigma = 0.7$, along with an adjustment achieved by multiplying the result by the sum of the squared weights (1.04). This adjustment is necessary because, with the roundoff error, the sum of the squared weights is not equal to one, and this property is required if we are to have local statistics that have standard normal distributions. The results yield a critical value of

$$y^* = \sqrt{-1.914 \ln\left(\frac{5.59\alpha}{m}\right)}$$ (7.24)

and a p-value of

$$p = \left(\frac{m}{5.59}\right) e^{-\frac{y_{max}^2}{1.914}}$$ (7.25)

It is also possible of course to look at larger kernel sizes associated with more smoothing, and this is desirable when searching for larger clusters. To take one more example, when $\sigma = 1.5$, a 7×7 filtering matrix is required (square filtering matrices that contain weights and are centered on cells are 1×1, 3×3, 5×5, 7×7, etc.) – beyond that range, weights are close to zero. Again, Figure 7.2 shows the distances of these cells from the focal cell, and Figure 7.3 shows the associated weights.

$$\begin{bmatrix} 0.01 & 0.03 & 0.06 & 0.08 & 0.06 & 0.03 & 0.01 \\ 0.03 & 0.10 & 0.19 & 0.23 & 0.19 & 0.10 & 0.03 \\ 0.06 & 0.19 & 0.36 & 0.45 & 0.36 & 0.19 & 0.06 \\ 0.08 & 0.23 & 0.45 & 0.56 & 0.45 & 0.23 & 0.08 \\ 0.06 & 0.19 & 0.36 & 0.45 & 0.36 & 0.19 & 0.06 \\ 0.03 & 0.10 & 0.19 & 0.23 & 0.19 & 0.10 & 0.03 \\ 0.01 & 0.03 & 0.06 & 0.08 & 0.06 & 0.03 & 0.01 \end{bmatrix}$$

Figure 7.3 Associated weights

Again using (7.12), the critical value is now

$$y^* = \sqrt{-1.77\ln\left(\frac{11.29\alpha}{m}\right)}$$ (7.26)

and the *p*-value associated with the observed statistic is

$$p = \left(\frac{m}{11.29}\right)e^{-\frac{y^2_{max}}{1.77}}$$ (7.27)

One noteworthy context for these scan statistics is that of the Getis–Ord statistic – discussed as a local statistic in Section 6.7. There will often be interest in assessing the significance of the maximum among all of the Getis–Ord statistics that can be defined for the many locations on the map. The scan statistics above allow for just this sort of assessment of the maximum Getis–Ord statistic in the special cases of data on a regular lattice with either (a) binary weights defined so that the rectangular scan statistics described in Section 7.6 can be used, or (b) Gaussian weights, which leads to an application of the Gaussian scan statistic.

7.6 Rectangular Scan Statistic

Siegmund and Yakir (2000) give results for a $r_1 \times r_2$ rectangular scanning window, where the weights used are rectangular (i.e., uniform), meaning that each observation in the window receives equal weight. Their statistic, in the case of normal variates, is the square of the sum of values,

$$S = \left(\sum_{\forall i \in R} y_i\right)^2$$ (7.28)

where the sum is taken for all locations i within the rectangle R. The probability of obtaining a higher value than the observed statistic (s), when the null hypothesis is true, is

$$\Pr(S > s) = \frac{4ma^{1.5}e^{-a}\prod_{i=1}^{2}e^{-1.166\sqrt{\frac{a}{r_i}}}}{\sqrt{\pi r_1 r_2}}$$ (7.29)

where m is the number of cells in the study region, and $a = s/(2r_1 r_2)$.

─────────(**Example**)─────────

Determine whether there is a significant clustering of z-scores in the following data, using σ = 0.7 and the interior 5 x 5 sub-matrix to avoid edge effects. Use 3 x 3 Gaussian (σ = 0.7) and 3 × 3 rectangular filters.

(Continued)

$$\begin{bmatrix} -0.91 & -0.54 & 0.95 & 0.60 & -0.74 & 1.20 & 0.47 \\ -1.70 & 0.88 & -0.80 & 0.74 & 0.05 & 0.17 & -0.27 \\ -0.25 & 0.45 & 0.36 & 1.12 & -0.51 & 2.10 & -0.13 \\ 2.02 & 1.41 & 0.55 & 0.56 & -1.45 & 0.85 & -1.20 \\ -0.05 & 1.49 & -0.82 & -1.36 & -0.11 & 1.13 & 1.00 \\ 0.03 & 1.29 & 1.07 & -0.82 & 1.39 & -1.47 & 0.32 \\ -0.19 & 0.15 & -1.17 & -0.92 & -1.84 & 0.73 & -0.68 \end{bmatrix}$$

Using the 3 x 3 Gaussian filter in Equation 7.23, the maximum local statistic is equal to 2.383, found when the filter is centered on the observation in the fifth row and second column of the data ($z = 1.49$). With $\alpha = 0.05$ the critical value is 2.93; the p-value is 0.23.

Using a 3 x 3 rectangular filter, the maximum sum is 6.99, and so the observed statistic is $s = 6.99^2 = 48.86$. Following Equation 7.29 the p-value is found to be 0.20.

Exercises

1. The number of cases observed and expected across six regions is as follows:

Region	Observed	Expected
1	10	6
2	15	7
3	8	8
4	9	10
5	10	12
6	12	21

Note that the total numbers for the observed and expected columns are equal. Use both the chi-square goodness-of-fit test and the likelihood ratio (G) statistic to test the null hypothesis that the observed counts could have arisen by chance, given the expected values.

2. Determine whether there is a significant clustering of z-scores in the following data, using $\sigma = 0.7$ and the interior 5x5 sub-matrix to avoid edge effects. Use 3x3 Gaussian ($\sigma = 0.7$) and 3x3 rectangular filters.

$$\begin{bmatrix} -0.81 & -0.64 & 1.05 & 0.62 & -0.79 & 1.10 & 0.87 \\ -1.53 & 0.97 & 1.20 & 0.94 & 0.76 & 0.67 & -0.17 \\ -0.22 & 0.48 & 0.92 & 1.92 & 0.71 & 2.20 & 1.13 \\ 1.01 & 0.46 & 0.85 & 0.89 & -0.14 & 0.85 & 1.20 \\ -0.35 & 1.94 & -0.82 & -1.36 & -0.11 & 1.13 & 1.00 \\ 0.39 & 1.11 & 1.45 & -0.48 & 1.33 & -1.27 & 0.72 \\ -0.49 & 0.19 & -1.47 & -0.22 & -1.44 & 0.93 & -0.98 \end{bmatrix}$$

3. A 5 x 5 grid contains the following values:

−1.82	−0.46	0.54	0.51	0.16
−0.68	0.25	−0.52	−0.60	−0.25
−0.93	1.71	0.57	−0.83	−0.67
1.21	1.14	0.34	−1.11	−0.06
−0.80	−1.29	−0.06	0.40	0.09

(a) Use the 3 x 3 Gaussian filter (Equation 7.23) to smooth the data. Use this filter on the 3 x 3 matrix of values contained at the center of the 5 x 5 grid of values above. You should have nine smoothed values – one at each of the nine locations at the center of the 5 x 5 grid.

(b) Compare the maximum of these nine smoothed values with the critical value given in Equation 7.24.

(c) Find the p-value, using Equation 7.25.

Further reading

The classic reference on scan statistics in the context of geographic patterns is the paper by Kulldorff and Nagarwalla (1995). Dale and Fortin (2014) also have a good discussion of scan statistics. The paper by Worsley (1996) provides a very accessible introduction to scan statistics that relies on principles of integral geometry and geometric probability.

Online resources

Visit **https://study.sagepub.com/rogersonspatialstatistics** for resources which supplement the material presented in this chapter. Students can find datasets and additional exercises, which can be used to practice the techniques covered in this chapter, as well as a chapter introduction video. The topic-software chart also describes the software packages that can be used to implement key techniques.

Lecturers and instructors can find exercises and conceptual prompts to facilitate classroom discussion and practice.

8

Spatial Means,
Spatial Models, and
Spatial Regression

┤Chapter overview├

In this chapter, you will learn about:

- Spatial means
- Some spatial models
- Estimation of autocorrelation parameter
- Type I and Type II errors, p-values, and critical values with spatial data
- Spatial regression

8.1 Spatial Means

When we have a number of observations in space, how do we find the mean of those observations? This is obviously a simple question, but it does not have a simple answer. Again the culprit is spatial autocorrelation – observations near to one another in space tend to be related. Thus to find the mean, we can't simply add the observed values and then divide by the number of observations. If two observations are near to one another in space, we do not have two independent observations. The amount of information contained in the two values is the equivalent of an effective number of independent observations that is somewhere between 1 and 2.

There are actually two related problems here. One is to find a good estimate of the mean that accounts for spatial dependence. The other problem is that our estimate of the variance of the estimate of the mean also needs to account for spatial dependence. If we incorrectly go ahead and assume independent observations, we will be overly confident in our estimate of the mean – if we were to construct confidence intervals in the usual way, they would be too narrow. With spatial data, where covariances are typically positive (high values tend to be near high values; low values tend to be near low values), the variance of the mean will be higher (and consequently, if we ignore spatial dependence when it exists, we will typically underestimate the variance of the mean).

Back in Chapter 1, we introduced this question by asking how to find the mean with three observations:

$$x_1 = 0.9889; x_2 = 1.388; x_3 = -0.6242 \qquad\qquad (8.1)$$

We looked briefly at two cases – one where the standard deviation is known, and one where it is estimated. If the standard deviation is known and $\sigma = 1$, a 95% confidence interval for the sample mean with known standard deviation is $0.5842 \pm 1.96\,1/\sqrt{3} = 0.5842 \pm 1.132$. If we don't know the standard deviation, $0.5842 \pm 4.303(1.065)/\sqrt{3} = 0.5842 \pm 2.646$ where 1.065 is the sample standard deviation, and where 4.303 is the value of a t-distribution with $n - 1 = 2$ degrees of freedom that leaves 2.5% of the area in each tail. For the hypothesis test that the true mean is 0, with known standard deviation, our z-score would be $0.5842/(1/\sqrt{3}) = 1.012$. With unknown and estimated standard deviation, $t = 0.5842/(1.065/\sqrt{3}) = 0.95$. In both cases, by comparing these values with the appropriate critical values, we would fail to reject the null hypothesis.

Our initial goals here are to (1) find an estimate of the mean that is better than the sample mean, which is equal to 0.5842, and (2) recognize, and correct for, the fact that by assuming independent observations, our estimate of the variance of the estimated mean is too low and our confidence in the estimated mean will be too high.

We begin by looking at the variance of the sample mean.

When you have spatial dependence, you have less confidence in the mean, because you have a smaller number of effectively independent observations.

One improvement is possible by recognizing that the usual estimate of the standard deviation of the mean, $\hat{\sigma}_{\bar{x}} = \sigma/\sqrt{n}$, is not appropriate when observations are not independent. When there is spatial dependence,

$$\sigma_{\bar{x}}^2 = \left(\frac{1}{n^2}\right)\Sigma_{i=1}^n V[x_i] + \left(\frac{2}{n^2}\right)\Sigma_{i<j} \mathrm{Cov}\left(x_i, x_j\right) \qquad\qquad (8.2)$$

where $V[x_i]$ is the variance of x_i (equal to σ_i^2), and $\mathrm{Cov}(x_i, x_j)$ is the covariance of observations i and j. When the covariances are zero, this reduces to the familiar $\hat{\sigma}_{\bar{x}}^2 = \sigma^2/n$. When the covariances are positive, as they often will be with spatial data, the variance of the sample mean becomes greater. If we ignore these covariances, our estimate of the variance of the sample mean will be too low – our confidence intervals will be too narrow, the absolute value of our test statistics will be too high (relative to the usual critical values), and our p-values will be too low. An important consequence is that, with spatial dependence, if we proceed with the usual statistical tests that assume independent observations, we will reject too many true null hypotheses – we will commit too many Type I errors.

In our example, if we are assuming that $\sigma_i^2 = 1$ for each of the three observations, and if we are given for example that $Cov(x_1, x_2) = Cov(x_2, x_1) = 0.8$ (the estimation of this covariance is a separate matter – and one which we will address later; for now we assume it is known, or given), then we have, from Equation 8.2, the following variance of the sample mean:

$$\sigma_{\bar{x}}^2 = \left(\frac{1}{9}\right)(1+1+1) + \left(\frac{2}{9}\right)(0.8) = \frac{3}{9} + \frac{1.6}{9} = \frac{4.6}{9} = 0.5111 \tag{8.3}$$

The standard error of the mean (equivalent to the standard deviation of the sampling distribution of the sample mean) is the square root of this, or 0.7149. Without dependence, the variance of the sample mean and the standard error are smaller. In particular, the second term on the right-hand side of the equation is zero, and $\sigma_{\bar{x}}^2 = 1/3$; the standard error is therefore $1/\sqrt{3} = 0.5774$. In the presence of spatial dependence, it would be a mistake to use a standard error of 0.5774. Use of the correct value of 0.7149 would give wider confidence intervals, and a smaller observed z-statistic. Our revised confidence interval, for the case of known variance (where we're assuming $\sigma = 1$), is now

$$0.5843 \pm 1.96(0.7149) = 0.5843 \pm 1.401 \tag{8.4}$$

Recall that, previously, the width of the confidence interval in this case was just 1.132.

We still have a number of questions. How should the mean be estimated, especially in light of our desire to down weight observations 1 and 2? What about the standard error associated with any new estimates for the mean? How do things change if the standard deviation and covariances are unknown?

The expression for the variance of the sample mean given in Equation (8.2) can be written in matrix form as

$$\sigma_{\bar{x}}^2 = \left(\frac{1}{n^2}\right)\sum_{i=1}^{n} V[x_i] + \left(\frac{2}{n^2}\right)\sum_{i<j} Cov(x_i, x_j) = \left(\frac{1}{n^2}\right)\sum_{i=1}^{n} V[x_i] + \left(\frac{1}{n^2}\right)\sum_{i=1}^{n}\sum_{j=1}^{n} Cov(x_i, x_j) \tag{8.5}$$

$$= (\mathbf{1}' \mathbf{V} \mathbf{1}) / n^2$$

where **1** and **1**′ are, respectively, column and row vectors each containing n ones, and **V** is the symmetric n × n matrix of variances and covariances of observations:

$$\mathbf{V} = \begin{bmatrix} \sigma_{11}^2 & \cdots & \sigma_{1n}^2 \\ \vdots & \ddots & \vdots \\ \sigma_{n1}^2 & \cdots & \sigma_{nn}^2 \end{bmatrix} \tag{8.6}$$

with the variances $\sigma_{ii}^2 = \sigma_i^2$ along the diagonal, and where the off-diagonal contains the covariances $(\text{Cov}(x_i, x_j) = \sigma_{ij}^2)$. See Appendix A for a review of matrix algebra.

As an aside, and/or reminder, the correlation ρ_{ij} is a standardized covariance, where the covariance of i and j is divided by the product of the standard deviations of i and j. Thus

$$\rho_{ij} = \frac{\sigma_{ij}}{\sigma_i \sigma_j} \tag{8.7}$$

The quantity $\mathbf{1}'\mathbf{V1}$ is simply the sum of the elements of the variance–covariance matrix. Continuing with our example, with $n = 3$ observations,

$$\mathbf{V} = \begin{bmatrix} 1 & 0.8 & 0 \\ 0.8 & 1 & 0 \\ 0 & 0 & 1 \end{bmatrix} \tag{8.8}$$

and $\mathbf{1}'\mathbf{V1} = 4.6$. The variance of the sample mean is thus $4.6/9 = 0.5111$, and the standard deviation is equal to the square root of this (0.7149), quantities we found earlier before introducing matrix notation.

The sample mean is unbiased, but it is not a minimum variance estimator. The maximum likelihood estimator of the mean is also unbiased, and the variance of the maximum likelihood estimator is lower than that of the sample mean, and so it is preferred, when data exhibit spatial dependence.

Maximizing the likelihood by setting the derivative equal to 0 yields the maximum likelihood estimator for the mean:

Haining (1988) gave the maximum likelihood estimator of the mean, μ, as

$$\tilde{\mu} = \frac{\mathbf{1}'\mathbf{V}^{-1}\mathbf{y}}{\mathbf{1}'\mathbf{V}^{-1}\mathbf{1}} \tag{8.9}$$

where $\mathbf{y}$ is the $n \times 1$ vector of observations. The reader can verify that the quantities in the numerator and denominator are scalars (e.g., 1×1). Furthermore, the denominator is equal to the sum of the elements of the inverse of $\mathbf{V}$. Notice that if the covariances are all equal to 0, $\mathbf{V} = \mathbf{I}$, and the expression reduces to the sample mean. In algebraic terms, the numerator is

$$\sum_{k=1}^{n} \left(\sum_{i=1}^{n} v_{ik}^{(-1)} \right) y_k \tag{8.10}$$

where $v_{ik}^{(-1)}$ is the element in row i, column k of the matrix $\mathbf{V}^{-1}$ and $\left(\sum_{i=1}^{n} v_{ik}^{(-1)} \right)$ is the sum of elements in column k of $\mathbf{V}^{-1}$. For the numerator then, we use the column sums of the inverse as weights associated with the observations.

In the present example V^{-1} is

$$V^{-1} = \begin{bmatrix} 2.78 & -2.22 & 0 \\ -2.22 & 2.78 & 0 \\ 0 & 0 & 1 \end{bmatrix} \tag{8.11}$$

and the three column sums that serve as weights for the observations are 0.56, 0.56, and 1, respectively. The reader may also wish to verify that indeed $V \, V^{-1} = I$. Completing the calculation of the numerator, the weighted sum of observations is $1'V^{-1}y = (0.56)$ $(0.9899) + (0.56)(1.388) + (1)(-0.6242) = 0.7074$.

We now want to divide this by a number less than 3, since we have less than three effectively independent observations. The quantity in the denominator of Equation 8.9, $1' \, V^{-1}1$, may be thought of as the effective sample size. Here $1' \, V^{-1}1 = 2.11$ (there is a small amount of roundoff error, because the elements of V^{-1} are only reported to two decimal places above). Equation 8.9 implies that the maximum likelihood (ML) estimator of the mean is $\tilde{\mu} = 0.7074 / 2.11 = 0.3353$.

To summarize, the numerator of Equation 8.9 is a sum of the products of the column sums of V^{-1} and the corresponding y-values:

$$(0.56)(0.9889) + (0.56)(1.388) + (1)(-0.6242) = 0.7074 \tag{8.12}$$

This is a weighted sum of the y-values, and the ML estimate of the mean is found by dividing this by the sum of the weights ($1' \, V^{-1}1 = 2.11$). The ML estimate is just a weighted average – found by weighting values and dividing by the sum of the weights. Note here in this example that the weights associated with the first two observations are smaller than 1; the high correlations between the first two observations leads us to "downweight" them in our calculation of the mean.

What about the variance of the estimate? If we were all to go out and take samples we'd all have different estimates of the mean. The distribution of estimates would be unbiased – it would be centered on the true mean. How variable would those estimates be? Haining (1988) gave the variance of this estimate as

$$\sigma_{\tilde{\mu}}^2 = \left(1' \, V^{-1} 1 \right)^{-1} \tag{8.13}$$

and in our case this is equal to $1/2.11 = 0.4737$. The square root of this is the standard deviation, and is equal to 0.6882. This makes intuitive sense – the variance is higher than in the case of independent observations ($\sigma^2/n = 1/3 = 0.333$) and we are simply using the effective sample size instead of the nominal (i.e., actual) sample size of $\sigma^2(1'V^{-1}1)^{-1} = \sigma^2/n'$, where n' is the effective sample size.

Also of note is the fact that this is less than the standard deviation associated with the sample mean (0.7149). The maximum likelihood estimator has slightly less uncertainty associated with it than does the simple average. The *relative efficiency* of $\bar{x}$ with respect to $\tilde{\mu}$ is given by the ratio of the variances:

$$(1' V^{-1}1)(1' V1)/n^2 \tag{8.14}$$

In this example, the relative efficiency is equal to $(4.6)(2.11)/9 = 1.078$, implying that the variance of the sample mean is 7.8% higher than the variance of the mean found via maximum likelihood.

8.2 Spatial Models

Often the covariances between pairs of observations are unknown and need to be estimated. We expect that, in many cases, the correlation between nearby observations will be high, but we don't know *how* high. We also don't know how these correlations decline or change with increasing distance between observations.

Rather than estimate all $n(n-1)/2$ separate pairs of covariances, it is common to construct a model for the covariance elements. A particularly widely used model is the spatial autoregressive (SAR) model (see, e.g., Anselin, 1988; Haining, 2003). It may be specified as

$$y_i = \mu + \rho \sum_{j=1}^{n} w_{ij} y_j + \varepsilon_i \tag{8.15}$$

where ρ is a measure of spatial dependence and w_{ij} is a weight reflecting the strength of the connection between locations i and j. ε_i is a disturbance (or error) term that describes the difference between the observed value of y and the modeled value of y; it has mean 0 and variance σ_i^2. With this model, values are functions of surrounding values. In matrix terms,

$$\mathbf{Y} = \mathbf{\mu} + \rho \mathbf{WY} + \varepsilon \tag{8.16}$$

where $\mathbf{Y}$ is an $n \times 1$ column vector of regional observations, $\mathbf{\mu}$ is an $n \times 1$ column vector with all elements equal to the overall mean, ρ is a scalar measure of spatial dependency, and $\mathbf{W}$ is an $n \times n$ weight matrix capturing the connectivity between regions. The $n \times 1$ column vector ε contains independent, normally distributed error terms all with mean 0. Equation 8.16 may be rewritten as

$$\mathbf{Y} - \rho \mathbf{WY} = \varepsilon \tag{8.17}$$

or

$$(\mathbf{I} - \rho \mathbf{W})\mathbf{Y} = \varepsilon \tag{8.18}$$

Premultiplying both sides by $(I - \rho W)^{-1}$,

$$Y = (I - \rho W)^{-1} \varepsilon \tag{8.19}$$

Letting $\sigma^2 I = V[\varepsilon]$, and using the fact that $\text{Var}[aX^2] = a^2 \text{Var}[X]$, the variance–covariance matrix associated with Y is

$$V[Y] = \sigma^2 (I - \rho W)^{-1} (I - \rho W')^{-1} = \sigma^2 \left[(I - \rho W') (I - \rho W) \right]^{-1} = \sigma^2 V \tag{8.20}$$

where $V = \left[(I - \rho W') (I - \rho W) \right]^{-1}$. The inverse of V has the simple form

$$V^{-1} = (I - \rho W') (I - \rho W) \tag{8.21}$$

The sum of the elements of this matrix (a useful quantity, as we have seen) is

$$n' = 1'V^{-1}1 = n - \rho \left[2(1'W1) - \rho (1'W^2 1) \right] = n - \rho \left[2\Sigma_1 - \rho \Sigma_2 \right] \tag{8.22}$$

where Σ_1 and Σ_2 are, respectively, the sum of the elements of W (i.e., $1'W1$) and W^2 (i.e., $1'WW1$). Haining (1988) provides a similar expression.

Note that in the special case where $\rho = 0$, an observation is modeled as the sum of the true mean μ and the normally distributed error term, ε, and the variance of an observation is equal to the variance of the error term. In the general case when ρ is not equal to 0, observations are functions of (a) the overall mean, (b) a normally distributed error term at that location, and (c) the normally distributed error terms at other locations.

The model is used here in part for expediency – it is a particularly simple spatial model both to simulate and to work with analytically. Note that the inverse of V has the simple form given in (8.21) and that the sum of the elements of V^{-1} can be written as in (8.22).

Aside: Simulation of values from a multivariate normal distribution

The following discussion follows closely that of Haining (1988: 116–17). To generate an $n \times 1$ column vector of values, Y, that have mean $\mu_{n \times 1}$ and variance–covariance matrix V, take

$$Y = \mu_{n \times 1} + L_{n \times n} e_{n \times 1} \tag{8.23}$$

where e is a column vector of independent, normally distributed variables with mean 0 and variance 1, and where L is found from the Cholesky decomposition of V:

$$V = LL' \tag{8.24}$$

L is an $n \times n$ lower triangular matrix (with entries of zero above the diagonal). From (8.19) and (8.20) we see that spatial autoregressive processes may be easily simulated by taking

$$L = (I - \rho W)^{-1} \tag{8.25}$$

and then using this matrix as a premultiplier of an $n \times 1$ vector of independent standard normal random variates.

Example

We begin by continuing with an example similar to the previous one, where there are three locations, and where Locations 1 and 2 are near one another, and both are far from Location 3. The **W** matrix reflects the strong connection between Location 1 and 2, and is specified as

$$W = \begin{bmatrix} 0 & 1 & 0 \\ 1 & 0 & 0 \\ 0 & 0 & 0 \end{bmatrix} \tag{8.26}$$

Convention is to have the diagonal of the **W** matrix set equal to 0.

Suppose it is known that $\rho = 0.6$ and $\sigma = 1$, and we observe

$$y' = [1.664 \; 0.7275 \; -0.0746]$$

We wish to find not only the sample mean and its standard error (i.e., the standard deviation of the mean), but also the maximum likelihood estimate of the mean and its associated standard error. The latter accounts for spatial dependence.

The sample mean is 0.7723. The standard deviation of the mean is $\sigma_{\bar{x}} = \sigma/\sqrt{n} = 1/\sqrt{3} = 0.577$.

To adjust these results for spatial dependence, it is perhaps easiest to first find the inverse of **V**:

$$V^{-1} = (I - \rho W)'(I - \rho W) = \begin{bmatrix} 1 & -0.6 & 0 \\ -0.6 & 1 & 0 \\ 0 & 0 & 1 \end{bmatrix} \begin{bmatrix} 1 & -0.6 & 0 \\ -0.6 & 1 & 0 \\ 0 & 0 & 1 \end{bmatrix} \tag{8.27}$$

$$= \begin{bmatrix} 1.36 & -1.20 & 0 \\ -1.20 & 1.36 & 0 \\ 0 & 0 & 1 \end{bmatrix}$$

The inverse of this is

$$V = \begin{bmatrix} 3.32 & 2.93 & 0 \\ 2.93 & 3.32 & 0 \\ 0 & 0 & 1 \end{bmatrix} \tag{8.28}$$

Then the corrected estimate of the standard deviation of the mean is

$$\sigma_{\bar{x}} = \sqrt{\frac{1'V1}{n^2}} = \sqrt{\frac{13.5}{9}} = 1.225 \tag{8.29}$$

and this is substantially higher than the common estimate (0.577) found above. For the maximum likelihood estimate of the mean, we have

$$\tilde{\mu} = \frac{1'V^{-1}y}{1'V^{-1}1} = \frac{0.308}{1.32} = 0.2334 \tag{8.30}$$

and note that the numerator is a weighted sum of the observations: $(0.16)(1.664) + (0.16)$ $(0.7275) + (1)(-0.0746) = 0.308$. Furthermore,

$$\sigma_{\tilde{\mu}} = \sqrt{\left(1'V^{-1}1\right)^{-1}} = \sqrt{\frac{1}{1.32}} = 0.8704 \tag{8.31}$$

and this is lower than the estimate of 1.225 found as the standard deviation of the sample mean – consistent with the result that maximum likelihood estimates are "better." Although both the ML estimate and the sample mean are unbiased, the ML estimate has a lower mean square error because its associated variance is lower.

8.3 Estimation of ρ

The spatial dependence parameter, ρ, may be estimated in a number of ways. When ρ is close to zero, Moran's I (Moran, 1950) is sometimes used, but this is not recommended (see, e.g., Ord, 1975). Ord (1975) showed that a closed-form estimate derived via least squares is

$$\hat{\rho} = \frac{z'Wz}{z'W'Wz} \tag{8.32}$$

but this is not a consistent estimator (a consistent estimator is one where the probability that the estimator converges to the true value as the sample size gets large converges to 1). The maximum likelihood estimator does not have a simple, closed-form solution. Alternatively, the APLE estimators given in the following section (Equations 8.34 and 8.37) may be used.

8.3.1 Estimation of ρ When μ is Known

Up until now, the spatial dependence parameter was taken as known, or given. In practice, this is rarely if ever the case.

If the data have been detrended to have mean 0 and we can assume that μ is known, the spatial dependence parameter can be estimated in a number of ways. Perhaps the simplest is the approximate profile likelihood estimator (APLE) of Li et al. (2007). They show that if $E[\mathbf{Y}] = 0$ (as it is for example with regression residuals), then

$$\hat{\rho} = \frac{\mathbf{Y}'\left(\dfrac{\mathbf{W}' + \mathbf{W}}{2}\right)\mathbf{Y}}{\mathbf{Y}'\mathbf{W}'\mathbf{W}\mathbf{Y} + \dfrac{\lambda'\lambda}{n}\mathbf{Y}'\mathbf{Y}} \tag{8.33}$$

where λ is the n x 1 vector of eigenvalues associated with $\mathbf{W}$. An alternative version that avoids the need to calculate eigenvalues is

$$\hat{\rho} = \frac{\mathbf{Y}'\left((\mathbf{W}' + \mathbf{W})/2\right)\mathbf{Y}}{\mathbf{Y}'\mathbf{W}'\mathbf{W}\mathbf{Y} + \text{Tr}(\mathbf{W}\mathbf{W})\mathbf{Y}'\mathbf{Y}/n} \tag{8.34}$$

where "Tr" indicates that the trace (equal to the sum of the diagonals) is to be taken. This estimator is an approximation to the maximum likelihood estimator – it is found by substituting the maximum likelihood estimator for the variance into the likelihood equation, differentiating with respect to ρ and setting the result equal to 0, and then taking a linear approximation to the resulting non-linear equation for ρ.

In the present example, where $\mathbf{y}' = [1.664 \ 0.7275 \ -0.0746]$, and $w_{12} = w_{21} = 1$ (with all other elements of $\mathbf{W}$ equal to zeros), this yields $\hat{\rho} = 0.4402$. We can use this estimate to go back and find the maximum likelihood estimate of σ^2 (where this is the variance associated with the data; note in the equation below that if $\rho = 0$, the variance calculation reduces to the usual sum of squared deviations from the mean (which is 0 in this case, since the data have been detrended), divided by n):

$$\hat{\sigma}^2 = \frac{1}{n}\mathbf{Y}'(\mathbf{I} - \rho\mathbf{W})'(\mathbf{I} - \rho\mathbf{W})\mathbf{Y} \tag{8.35}$$

In this example, we have $\hat{\sigma}^2 = 0.6038$ and $\hat{\sigma} = \sqrt{0.6038} = 0.7770$. This may be compared with the "usual" calculation of the variance where spatial dependence is ignored, which would yield 0.5048 (and $\hat{\sigma} = 0.7105$) if $n = 3$ was used in the denominator.

This last calculation serves as a reminder that the maximum likelihood estimator of the variance is biased; we know from first-semester statistics that as a result, $n - 1$ is used instead of n in the denominator. In our case, using $n = 2$ in the denominator results in a variance of $(3/2)(0.5048) = 0.7572$ when ignoring spatial dependence, and $(0.6038)(3/2) = 0.9057$ when accounting for it.

For the standard deviation associated with the maximum likelihood estimator of the mean, we have

$$\sigma_{\hat{\mu}} = \sqrt{0.9057\left(\mathbf{1}'\mathbf{V}^{-1}\mathbf{1}\right)^{-1}} = \sqrt{\frac{0.9057}{1.32}} = 0.8283 \tag{8.36}$$

8.3.2 What Happens When μ, σ, and ρ Are All Unknown?

Li et al. (2012) give a version of their APLE estimator for the case where **Y** is a linear function of explanatory variables, and the regression parameters are unknown (for a brief treatment of ordinary least squares regression using matrix algebra, see Section A.3.4 in Appendix A). This includes a special case where there are no explanatory variables, and the mean is unknown.

In this case,

$$\hat{\rho} = \frac{\mathbf{Y}'(\mathbf{I}-\mathbf{P})((\mathbf{W}'+\mathbf{W})/2)(\mathbf{I}-\mathbf{P})\mathbf{Y}}{\mathbf{Y}'(\mathbf{I}-\mathbf{P})\mathbf{W}'\mathbf{W}\mathbf{Y}-\mathbf{Y}'(\mathbf{I}-\mathbf{P})(\mathbf{W}'+\mathbf{W})\mathbf{P}(\mathbf{W}'+\mathbf{W})(\mathbf{I}-\mathbf{P})\mathbf{Y}}$$
$$+\{\mathrm{Tr}(\mathbf{W}\mathbf{W})\}\mathbf{Y}'(\mathbf{I}-\mathbf{P})\mathbf{Y}/n \tag{8.37}$$

where $\mathbf{P} = \mathbf{X}(\mathbf{X}'\mathbf{X})^{-1}\mathbf{X}'$, and where X is a matrix that has n rows and $p + 1$ columns, where p is the number of explanatory variables, and where the first column has all elements equal to one. In the special case where there are no explanatory variables, X is just an $n \times 1$ column of ones (as is the case in the matrix form of linear regression, where the first column is a column of ones associated with the intercept). In this special case, **P** is an $n \times n$ matrix with all entries equal to $1/n$.

Although this is certainly a long expression, it still has the considerable advantages of (a) being a direct estimate of the spatial dependence parameter, where ρ does not have to be solved for iteratively, as is the case where no approximation to the likelihood equation is made (in that case, it is not possible to solve directly for ρ, by isolating it on the left side of the equation), and (b) as we will now see, the estimate can be used to find estimates of the mean and variance, with no need for further iterations.

In this example, we find $\hat{\rho} = -0.0940$. Note that the estimate is negative; we haven't constrained it to be non-negative. Using the estimate to find an estimate of the mean, we have

$$\tilde{\mu} = \frac{\mathbf{1}'\mathbf{V}^{-1}\mathbf{y}}{\mathbf{1}'\mathbf{V}^{-1}\mathbf{1}} = 0.8214 \tag{8.38}$$

This same result could also be found via the expression for parameter estimation using generalized least regression:

$$\hat{\beta} = (\mathbf{X}'\mathbf{V}^{-1}\mathbf{X})^{-1}\mathbf{X}'\mathbf{V}^{-1}\mathbf{Y} \tag{8.39}$$

where the reader will note that these last two equations are identical when X is an $n \times 1$ vector of ones.

For the variance, we have

$$\hat{\sigma}^2 = \frac{1}{n}(\mathbf{Y}-\tilde{\mu})'(\mathbf{I}-\rho\mathbf{W})'(\mathbf{I}-\rho\mathbf{W})(\mathbf{Y}-\tilde{\mu}) = \frac{1}{n}(\mathbf{Y}-\tilde{\mu})'\mathbf{V}^{-1}(\mathbf{Y}-\tilde{\mu}) \tag{8.40}$$

which in this case is equal to 0.4994, and this implies a standard deviation of $\hat{\sigma} = \sqrt{0.4994} = 0.7067$.

An alternate, more arduous route to a similar answer would be to iterate: find ρ assuming that $\mu = 0$ (Equation 8.34), then find the maximum likelihood estimate of the mean using Equation 8.8. Next use this value of the mean to update the estimate of ρ:

$$\hat{\rho} = \frac{(\mathbf{Y} - \tilde{\mu})'((\mathbf{W}' + \mathbf{W})/2)(\mathbf{Y} - \tilde{\mu})}{(\mathbf{Y} - \tilde{\mu})'\mathbf{W}'\mathbf{W}(\mathbf{Y} - \tilde{\mu}) + \lambda'\lambda / n(\mathbf{Y} - \tilde{\mu})'(\mathbf{Y} - \tilde{\mu})} \tag{8.41}$$

which in turn is used to update the estimate of $\tilde{\mu}$, and so on. This iterative scheme is close to converging at $\tilde{\mu} = 0.8191$ and $\tilde{\rho} = -0.0892$. We note here that this example is illustrative – it is not generally a good idea to try to estimate three parameters using just three observations!

Example

The distribution of $\hat{\rho}$

How good is the estimate of the spatial dependence parameter derived in this way? Li et al. (2007, 2012) gave examples that show that its performance (with respect to bias and mean square error) is similar to the maximum likelihood estimator, especially when the amount of spatial dependence is not too large. Here we will report the results of just one simulation exercise.

A 10 x 10 grid of square cells was used, with weights defined using rook's adjacency (pairs of cells sharing a side receive a 1 in the weight matrix), and then $\mathbf{W}$ was standardized to have row sums equal to one. An SAR process was simulated (with $\mu = 0$ and $\sigma = 1$) for four separate scenarios, corresponding to the following values for ρ: 0.1, 0.2, 0.4, and 0.6. For each of these scenarios, two variants were simulated; in one case, μ was assumed known (prompting use of Equation 8.34), and in the other case it was not (and so Equation 8.37 was used). For each of these cases, the process was simulated 500 times. The results are shown in Table 8.1. It shows that the estimate of ρ becomes a bit biased as the amount of dependence increases, underestimating the true value. It is interesting that the standard deviation of the estimate decreases with increasing dependence. This is also seen in Li et al.'s (2007) simulations (see, e.g., their Figure 1). Li et al. (2012) also gave an asymptotic expression for the variance of the APLE estimator.

Table 8.1 Mean and standard deviation of estimates of autocorrelation parameter

| | Estimates of ρ | | | |
| | μ known | | μ unknown | |
True ρ	$\hat{\rho}$	Standard deviation	$\hat{\rho}$	Standard deviation
0.1	0.0982	0.130	0.0821	0.138
0.2	0.1960	0.117	0.1753	0.130
0.4	0.3894	0.105	0.3604	0.117
0.6	0.5562	0.085	0.5343	0.089

The fairly large standard deviation when dependence is low also potentially raises a question. Moran's I is generally not used to estimate ρ, but its associated standard error is relatively smaller than that for the APLE estimator when ρ is small. It would be interesting to carry out a more detailed study – for example, comparing the mean square errors (equal to the variance plus the square of the bias) of APLE, maximum likelihood, and Moran's I estimators.

Relative Efficiency of Simple Average and Maximum Likelihood Estimators of the Mean

With spatial dependence, we have seen that there are at least two ways to estimate the mean and its standard error. We can use the simple average or we can use the maximum likelihood estimator; each has associated with it a standard error (Equations 8.2 and 8.13). The relative efficiency of these estimators (Equation 8.14) is often close to one. For example, with an $n \times n$ set of cells with rook's adjacency and row standardization, and with $\rho = 0.4$, the relative efficiency of the simple average with respect to the maximum likelihood estimator is 1.011, 1.004, and 1.002, for $n = 3$, 5, and 10, respectively. When ρ is increased to 0.8, these three values increase a bit, to 1.041, 1.022, and 1.009, but they are still fairly close to 1.

Although there is not a large difference between the two estimators when spatial dependence is small, the main lesson here is that it is important to do *something* instead of ignoring potential dependence entirely by simply calculating the average and using $\sigma/\sqrt{n}$ as the estimate for the standard error of the mean.

8.4 Type I and Type II Errors, *p*-Values, and Critical Values with Spatial Data

It is well known that geographic data are characterized by spatial dependence, and that this has the consequence of reducing the number of effectively independent observations. Therefore, either statistical tests must be adjusted – confidence intervals should be wider, and *p*-values and critical values should be higher – or spatial methods accounting for the dependence should be employed. Unadjusted analyses will have higher probabilities of Type I errors. The necessary adjustments, corrections, and calculations are not difficult to carry out if the amount of dependence is known or can be estimated, when simultaneous autoregressive and conditional autoregressive models are adopted. Of course if the amount of dependence is known, spatial methods can be used directly, but it is of interest to know the magnitude of the effect brought about by ignoring the dependence. Recent advances have made the former model particularly simple to estimate. Some of the results here in this section are simplified further by using a tight bound for the cumulative standard normal distribution. Wheat yield data from the Rothamsted Experimental Station will also be used for illustration.

When data are independent, standard methods of statistical inference may be applied to test hypotheses. Typically the analyst chooses a level of α, the probability of making a Type I error by incorrectly rejecting a true null hypothesis (common choices for α are 0.01, 0.05, and 0.10). The value of α is chosen prior to data collection and analysis, and although the researcher has control over it, it should not be made too small. This is because, as α declines, it becomes increasingly difficult to reject null hypotheses, even when they are false. There is an increasing probability of making a Type II error, where one incorrectly accepts a false null hypothesis.

It is well known that the probability of making these errors increases when data are not independent, as is often the case with spatial data. In particular, spatial dependence reduces the number of effectively independent observations (see, e.g., Griffith, 2005), since there is redundancy in the information collected at locations near to one another.

In this section, we assess the effects of failing to account for spatial dependence on p-values, critical values, and the probabilities of Type I and Type II errors. This is done first for spatial models where the amount of dependence is known. When the amount of dependence is unknown, it can be estimated, and the results serve as estimates of the effects that would be observed if the dependence were ignored. Of course if the amount of spatial dependence is either known or estimated, one can directly employ spatial models that account for that dependence. However an alternative would be to apply the traditional, usual methods, with proper adjustments (of, e.g., critical values) for the dependence. In addition, it would be interesting and useful to know and report the results of the magnitude of errors that *would* result, if the dependence were ignored.

The results (a) can follow, if desired, from a straightforward, one-step, closed-form estimator of the spatial dependence parameter (more accurate, though more complex, estimators could also of course be used), (b) make use of a tight bound for the cumulative normal distribution, if direct calculation (rather than the use of tables) is desired, and (c) treat the effects on Type II errors, which are often ignored.

Here we use specific models for spatial dependence (the spatial autoregressive model and the conditional autoregressive model). This allows use of known results to provide simple and easy-to-use equations for corrected critical values, p-values, and confidence intervals, and for the probability of Type I and Type II errors for hypothesis tests regarding means and regression parameters, when dependence is ignored.

After discussing the variance of the sample mean and effective sample size in the remainder of this section, the effects of spatial dependence on Type I errors, p-values, and critical values for one-sample tests about the mean are discussed in the following section. The remaining sections include a focus upon Type II errors, spatial models, generalizations, an application to the well-known dataset on wheat yields in Rothamsted, and a summary.

The presence of positive spatial autocorrelation inflates the variance of the sample mean. If spatial autocorrelation is ignored, it is incorrectly presumed that the variance is smaller than it really is, and confidence intervals around the sample mean are narrower than they should be. Correspondingly, for a null hypothesis regarding the true mean or differences in means, critical values of the test statistic and *p*-values will be too low, and the hypothesis might be rejected, when the correct decision would be to fail to reject it.

We have seen that the quantity $\mathbf{1}'\mathbf{V}^{-1}\mathbf{1}$, the sum of the elements of the inverse of $\mathbf{V}$, is, conceptually, the effective sample size (when the variance of all individual observations is equal to one). This quantity, denoted here as n', can be thought of as the number of effectively independent observations, and this declines as the amount of spatial dependency increases. Confidence intervals, *p*-values, critical values for test statistics, and the likelihood of Type I and Type II errors should all be based upon n' instead of n.

Now let

$$\psi = \frac{n'}{n} \tag{8.42}$$

The quantity Φ is a fraction between 0 and 1 that may be interpreted as a reduction factor acting on the number of original observations. A value of 2/3 for instance would imply that spatial dependence has effectively reduced the number of observations to a size equal to 2/3 of the original sample size. A key observation is that this quantity is also equal to the ratio of the variances of the sample means, comparing the case of spatial dependence with the (possibly assumed) case of no spatial dependence. The square root of Φ is equal to the ratio of the standard deviations of the sample means and it is a measure of the relative dispersion of the two distributions in Figure 8.1.

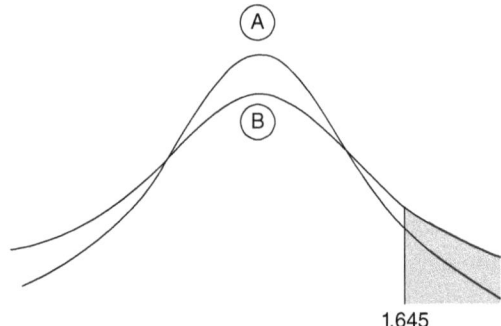

1.645

Figure 8.1   Distributions of the sample mean (standardized) with (B) and without (A) spatial dependence

8.4.1 Confidence Intervals, Type I errors, and Adjusted Critical Values and *p*-Values When Observations are Spatially Dependent

We begin by assuming that σ and $\mathbf{V}$ are known; this will not usually be the case, but discussion of this scenario easily reveals the effects of known forms of spatial dependence on *p*-values, critical values, and the probabilities of Type I and Type II errors. Discussion of the more usual situation where σ and V are unknown follows.

Suppose that we are considering the null hypothesis that the true mean (μ_0) is equal to zero, against the one-sided alternative hypothesis that the mean is greater than zero, with α = 0.05. Then the "usual" approach to hypothesis testing is to ignore any spatial dependence and to reject the null hypothesis when the observed standardized observation exceeds the critical value found from the standard normal distribution (assuming *n* is large):

$$z = \frac{\bar{x} - \mu_0}{s / \sqrt{n}} = \sqrt{n}\bar{x} / s > z_{crit} = 1.645 \tag{8.43}$$

More generally, for any choice of α, the critical value of z is determined as $\Phi^{-1}(1-\alpha)$, where this notation refers to the value of the standard normal distribution that leaves an area of α in the right-hand tail. For a one-sided alternative with the mean *less* than zero, the critical value is $\Phi^{-1}(\alpha)$.

But with spatial dependence we need to account for the fact that the effective sample size is lower (n') and the estimate of the standard deviation of the sample mean is higher ($s/\sqrt{n'}$ instead of $s/\sqrt{n}$). Using the maximum likelihood estimate of the variance of the mean (Equation 8.13), we reject the null hypothesis when $\sqrt{1'V^{-1}1}\bar{x} / s = \sqrt{n}\bar{x} / s > 1.645$. With this modified approach the desired Type I error level is maintained; if dependence is ignored, there are too many Type I errors. In Figure 8.1, distribution A is the distribution of the sample mean when all observations are independent. Distribution B is characterized by the larger variance that is associated with the distribution of sample means when data exhibit dependence. The value of 1.645 for distribution A has an area to the right that is equal to 0.05. The shaded area under distribution B is the actual Type I error; it corresponds to the area to the right of $1.645\sqrt{\psi}$. That is,

$$\alpha' = 1 - \Phi\left(1.645\sqrt{\psi}\right) \tag{8.44}$$

where $\Phi(\cdot)$ signifies the cumulative standard normal distribution.

Using a tight lower bound for the cumulative normal distribution (Sasvári and Chen, 1999), a lower bound for the Type I error that also serves as a good approximation for it is

$$\alpha' \approx \frac{1 - \sqrt{1 - e^{-\frac{2\left(1.645\sqrt{\psi}\right)^2}{\pi}}}}{2} \tag{8.45}$$

An alternative perspective is to ask how the critical value of the test statistic should be adjusted if the usual approach to hypothesis testing is taken. In the previous example, the use of 1.645 as a critical value would lead to the rejection of too many true null hypotheses and it would instead be desirable to use a higher critical value in its place. Conceptually, we wish to increase the critical value by the ratio of the standard deviations. Thus the critical value that would yield a Type I error probability of 0.05 is $1.645/\sqrt{\psi} = 1.645/\sqrt{n'/n}$. More generally, the corrected critical value for a one-tailed test (with the alternative mean greater than the mean associated with the null hypothesis) with desired Type I error equal to α may be expressed as

$$z'_{crit} = \frac{\Phi^{-1}(1-\alpha)}{\sqrt{\psi}} = z_{crit} / \sqrt{\psi} \qquad (8.46)$$

(For a one-sided alternative where the true mean is *less* than the hypothesized mean, $\Phi^{-1}(1-\alpha)$ is replaced with $\Phi^{-1}(\alpha)$.)

The *p*-value associated with a test statistic gives the probability of observing a more extreme result than the one observed, under the condition that the null hypothesis is true. This is underestimated when spatial dependency is not taken into account. The corrected *p*-value is equal to

$$p' = p / \sqrt{\psi} \qquad (8.47)$$

8.4.2 Type II Errors

Introductory treatments of statistical methods typically devote much less space to discussion of the probability of Type II errors (β), where false null hypotheses are accepted.

For a one-sided hypothesis about a mean, the situation is as in Figure 8.2. The distribution on the left is that of the sample mean (standardized by first subtracting the hypothesized mean (μ_0), and then dividing by $\sigma/\sqrt{n}$) when the null hypothesis is true. The alternative distribution on the right is that of the sample mean (again standardized), when the true mean (μ_1) is greater than the hypothesized mean, and where the difference between the two means is $\delta = \mu_1 - \mu_0$. The figure assumes that $\alpha = 0.05$ and that the critical value is 1.645. The area to the right of 1.645 on the null distribution that lies under the alternative distribution is equal to the power of the test ($1 - \beta$), since it is where the null hypothesis will be rejected when it is false. The shaded area to the left of 1.645 on the null distribution that is under the alternative distribution is equal to the probability of a Type II error (β), where the null hypothesis is incorrectly accepted.

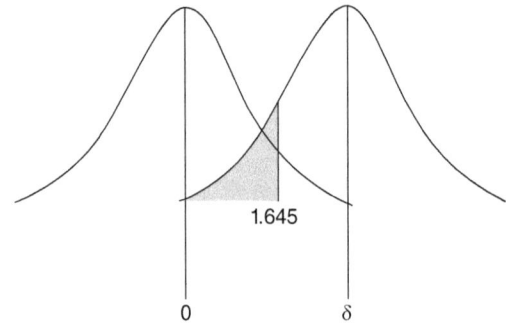

Figure 8.2 Type II errors

Note: The null distribution is on the left and the alternative distribution is on the right. Distributions have been standardized

The power of the test may be calculated by recognizing that the value of 1.645 on the null distribution corresponds to a value of $1.645 - \delta\sqrt{n}/\sigma$ on the alternative distribution. Using the cumulative distribution for the standard normal distribution, when there is no spatial dependence, or none is assumed, the calculated probability of a Type II error is

$$\beta = \Phi\left(1.645 - \delta\sqrt{n}/\sigma\right) \tag{8.48}$$

and the power of the test (i.e., the probability that a false null hypothesis is rejected) is equal to its complement

$$Power = 1 - \beta = 1 - \Phi\left(1.645 - \delta\sqrt{n}/\sigma\right) \tag{8.49}$$

With spatial dependence, use of the usual critical value will give a misleading overestimate of power, and an underestimate of β. While the usual calculation would be based on the solid lines in Figure 8.3, spatial dependence gives rise to the dashed curve in the figure when the null hypothesis is not true. Use of the usual critical value (e.g., 1.645) would lead to an actual Type II error of $\Phi\left(1.645 - \delta\sqrt{n'}/\sigma\right)$.

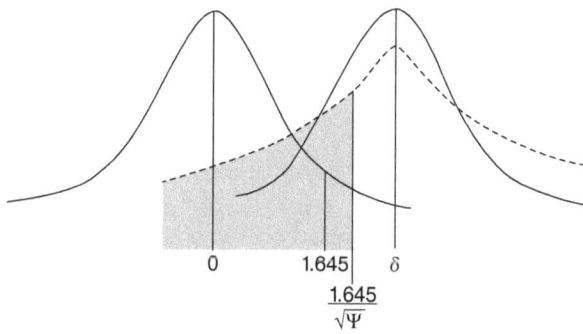

Figure 8.3   Type II errors with spatial dependence

Note: The alternative distribution is on the right

To account for spatial dependence, we have seen in the previous section that we should actually use a corrected critical value of $1.645/\sqrt{\psi}$. In this case, the corrected Type II error for the test is represented by the shaded area in the figure and this may be expressed as

$$\beta' = \Phi\left(1.645/\sqrt{\psi} - \delta\sqrt{n'}/\sigma\right) \tag{8.50}$$

Using the approximation for the cumulative standard normal distribution, this corrected Type II error is approximately equal to

$$\beta' \approx \frac{1 + \sqrt{1 - e^{-\left(\frac{2}{\pi}\right)\left(\frac{1.645}{\sqrt{\psi}} - \frac{\delta\sqrt{n'}}{\sigma}\right)^2}}}{2} \tag{8.51}$$

We next use the spatial autoregressive (SAR) model described in Section 8.4 and the conditional autoregressive model described here to illustrate these ideas; use of these models yield results for the quantity ψ that are easy to apply in practice.

The conditional autoregressive (CAR) model takes the value z_i at location i, *given* the values of z at other locations are normally distributed, and it has the following variance–covariance matrix:

$$V = (I - \rho W)^{-1} \tag{8.52}$$

For the SAR model, note that

$$\psi = \frac{n'}{n} = 1 - \frac{2\rho\Sigma_1}{n} + \frac{\rho^2\Sigma_2}{n} = 1 - \left(\frac{\rho}{n}\right)(2\Sigma_1 - \rho\Sigma_2) \tag{8.53}$$

where Σ_1 and Σ_2 are defined as the sum of W and W^2, respectively.
For the conditional autoregressive (CAR) model,

$$V = (I - \rho W)^{-1}; V^{-1} = (I - \rho W) \tag{8.54}$$

and the effective sample size is

$$n' = 1'V^{-1}1 = n - \rho(1'W1) = n - \rho\Sigma_1 \tag{8.55}$$

implying that

$$\psi = \frac{n'}{n} = 1 - \frac{\rho\Sigma_1}{n} \tag{8.56}$$

For row-standardized weight matrices based on binary adjacency, $\Sigma_1 = \Sigma_2 = n$, implying that, for the SAR model,

$$\psi = \frac{n'}{n} = (1 - \rho)^2 \tag{8.57}$$

$$\alpha' = 1 - \Phi[(1 - \rho)z_{crit}] \tag{8.58}$$

$$\alpha' \approx \frac{1 - \sqrt{1 - e^{-\frac{2(z_{crit}(1-\rho))^2}{\pi}}}}{2} \tag{8.59}$$

$$z'_{crit} = \frac{\Phi^{-1}(1-\alpha)}{1-\rho} \tag{8.60}$$

$$\rho' = \frac{\rho}{1-\rho} \tag{8.61}$$

where z_{crit} is the critical value of the standard normal distribution that is used for hypothesis tests with a chosen Type I error level of α, in the absence of spatial dependence.

For the CAR model, where weight matrices are row standardized,

$$\psi = \frac{n'}{n} = (1-\rho)$$ (8.62)

$$\alpha' = 1 - \Phi[1.645 \sqrt{(1-\rho)}\,]$$ (8.63)

$$\alpha' \approx \frac{1 - \sqrt{1 - e^{-\frac{2(1-\rho)z_{crit}^2}{\pi}}}}{2}$$ (8.64)

$$z^* = \Phi^{-1}(\alpha) / \sqrt{(1-\rho)}$$ (8.65)

$$p' = p / \sqrt{(1-\rho)}$$ (8.66)

As an example, we first take the simple case where we have a systematic spatial sample of $n = 9$ observations (Figure 8.4a). From these observations, we construct a binary weights matrix, **W** (Figure 8.4b). This matrix has entries of one for those pairs of locations that are adjacent to one another (using the definition of rook's adjacency), and zeros elsewhere. **W** is symmetric and has a total of 24 ones.

1	2	3
4	5	6
7	8	9

Figure 8.4a   Hypothetical nine-region system

$$\begin{bmatrix}
0 & 1 & 0 & 1 & 0 & 0 & 0 & 0 & 0 \\
1 & 0 & 1 & 0 & 1 & 0 & 0 & 0 & 0 \\
0 & 1 & 0 & 0 & 0 & 1 & 0 & 0 & 0 \\
1 & 0 & 0 & 0 & 1 & 0 & 1 & 0 & 0 \\
0 & 1 & 0 & 1 & 0 & 1 & 0 & 1 & 0 \\
0 & 0 & 1 & 0 & 1 & 0 & 0 & 0 & 1 \\
0 & 0 & 0 & 1 & 0 & 0 & 0 & 1 & 0 \\
0 & 0 & 0 & 0 & 1 & 0 & 1 & 0 & 0 \\
0 & 0 & 0 & 0 & 0 & 1 & 0 & 1 & 0
\end{bmatrix}$$

Figure 8.4b   Weight matrix for hypothetical nine-region system

In this example, with $n = 9$, $\sigma = 1$, and $\rho = 0.1$, for the SAR model with binary, rook's adjacency (and no row standardization) we have, using (8.53),

$$n' = n - \rho(2\Sigma_1 - \Sigma_2) = 9 - 0.1\left[2(24) - 0.1(68)\right] = 9 - 4.12 = 4.88 \tag{8.67}$$

Instead of having nine independent observations, there are effectively the equivalent of just under five independent observations. Table 8.2 shows how the effective sample size and the standard deviation of the sample mean decline with increasing spatial dependence. Confidence intervals for the sample mean $\bar{x}$ should use the sample standard deviation (s) and make use of n' instead of n. For example, a 95% confidence interval would be found as

$$\bar{x} \pm 1.96 \frac{s}{\sqrt{n'}} \tag{8.68}$$

For the SAR model with $n = 9$, $n' = 4.88$, $\sigma = 1$, $\rho = 0.1$, and $\sqrt{\psi} = \sqrt{4.88/9} = 0.7364$, the actual Type I error is found from Equation 8.44 as $\alpha' = 1 - \Phi(1.645\sqrt{\psi}) = 1 - \Phi(1.211) = 0.1129$.

Table 8.2 Effects of spatial dependence on effective sample size and standard deviation of sample mean (SAR model)

ρ	$\sigma_{\bar{x}} = \sigma/\sqrt{n}$	$\sigma_{\bar{\mu}} = \sigma/\sqrt{n'}$	$n' = 1'V^{-1}1$
0.00	0.333	0.333	9
0.05	0.385	0.384	6.77
0.10	0.458	0.453	4.88
0.15	0.568	0.548	3.33
0.20	0.750	0.687	2.12
0.24	1.011	0.846	1.40

Note: $n = 9$; $\sigma = 1$

Standardized weights

ρ	$\sigma_{\bar{x}} = \sigma/\sqrt{n}$	$\sigma_{\bar{\mu}} = \sigma/\sqrt{n'}$	$n' = 1'V^{-1}1$
0.00	0.333	0.333	9
0.1	0.371	0.370	7.29
0.2	0.417	0.417	5.76
0.3	0.478	0.476	4.41
0.4	0.559	0.556	3.24
0.5	0.672	0.667	2.25

Note: $n = 9$; $\sigma = 1$

The approximation and the lower bound to the Type I error (Equation 8.59) is equal to 0.1105, close to the more accurate value of 0.1129 found above using the standard normal table. The corrected critical value (Equation 8.60) is equal to $1.645/\sqrt{4.88/9} = 2.234$.

In this example, with $n = 9$ and $\sigma = 1$, suppose that the difference between the hypothesized and true values of the mean δ is equal to 1. Then with $\alpha = 0.05$, using Equation 8.48, $\beta = 0.0877$ and the power of the test is equal to $1 - 0.0877 = 0.9123$.

Use of the usual critical value with a correction for the effective sample size would lead to a Type II error of

$$\beta = \Phi\left(1.645 - \frac{\delta\sqrt{n'}}{\sigma}\right) = \Phi\left(1.645 - \frac{1\sqrt{4.88}}{1}\right) = 0.2864 \tag{8.69}$$

From Equation 8.50, the corrected Type II error is

$$\Phi\left(\frac{1.645}{\sqrt{\psi}} - \frac{\delta\sqrt{n'}}{\sigma}\right) = \Phi\left(\frac{1.645}{\sqrt{\frac{4.88}{9}}} - \frac{1\sqrt{4.88}}{1}\right) = \Phi\left(2.234 - \frac{1\sqrt{4.88}}{1}\right) = 0.5099 \tag{8.70}$$

For the SAR model with row-standardized weights, an approximation to the corrected Type II error is (Equation 8.51):

$$\beta' \approx \frac{1 + \sqrt{1 - e^{-\left(\frac{2}{\pi}\right)\left(\frac{1.645}{(1-\rho)} - \frac{\delta\sqrt{n'}(1-\rho)}{\sigma}\right)^2}}}{2} \tag{8.71}$$

and for the current example this is equal to 0.510.

8.4.3 m x m Systems

For a 3×3 regional system with rook's adjacency, we have seen that $\Sigma_1 = 24$ and $\Sigma_2 = 68$. What are the corresponding values in a general $m \times m$ regional system with rook's adjacency? The sum of the weights matrix, $\Sigma_1 = 1'\,W1$, can be found by recognizing that there are three types of subregions – (a) those occupying the corners, (b) those located on the sides, but not in the corners, and (c) those located in the "middle" – defined by not being on the sides or in the corners. The number of cells of each type, along with the number of subregions a cell of each type is connected to, is given in the top panel of Table 8.3. Summing the products of the entries in each row of the table yields the desired quantity

$$\Sigma_1 = 1'\,W1 = 4m(m-1) \tag{8.72}$$

Table 8.3   Connectivity in an $m \times m$ regional system (rook's adjacency)

	Type	Number of cells of this type	Number of subregions that a cell of this type is connected to via one step
(a)	Corner	4	2
(b)	Side; not corner	$4(n - 2)$	3
(c)	Middle (not side or corner)	$n^2 - 4n - 4$	4

	Type	Number of cells of this type	Number of subregions that a cell of this type is connected to via two steps
(a)	Corner	4	4
(b)	Side; adjacent to corner	8	6
(c)	Side; not adjacent to corner	$4(n - 4)$	7
(d)	Inner cells ("middle")		
	Adjacent to edge and diagonal to corner	4	10
	Adjacent to edge and not diagonal to corner	$4(n - 4)$	11
	Others	$n^2 - 8(n - 2)$	12

To find a general expression for Σ_2, the subregions of the $m \times m$ system are categorized as in the bottom panel of Table 8.3. The table gives (a) the number of cells of each type, and (b) for each type, the number of two-step connections to other regions. Summing the products of the entries in each row of the table yields

$$12(m - 1)^2 - 4 \tag{8.73}$$

We must also recognize that there are two-step connections that begin and end at each subregion – these are non-zero elements that constitute the diagonal of $\mathbf{W}^2$. There are Σ_1 such connections. Adding the two quantities yields the result

$$\Sigma_2 = \mathbf{1}' \, \mathbf{W}^2 \mathbf{1} = 12(m - 1)^2 - 4 + \Sigma_1 = 4(4m^2 - 7m + 2) \tag{8.74}$$

For large $m \times m$ square lattices with rook's adjacency, the above imply that ignoring edge effects leads to the approximation $\Sigma_1 \approx 4m^2$ and $\Sigma_2 \approx 16m^2$. In this case, for the SAR model,

$$\psi = \frac{n'}{n} = 1 - 8\rho + 16\rho^2 \qquad (8.75)$$

For example, when $\rho = 0.1$, $\psi = 0.36$. Note that when edge effects are ignored this quantity is not a function of m. For a hexagonal lattice, ignoring edge effects,

$$\psi = \frac{n'}{n} = 1 - 12\rho + 36\rho^2 \qquad (8.76)$$

8.4.4 Two-sample Tests

When the null hypothesis of no difference in two population means is tested, the standard deviation of the difference in sample means is, under the assumptions of homoscedasticity (i.e., equal variances for the two populations) and independent observations,

$$\sigma_{\bar{x}_1 - \bar{x}_2} = \sigma \sqrt{\frac{1}{n_1} + \frac{1}{n_2}} \qquad (8.77)$$

Again, this standard deviation will be higher when there is spatial dependence in the two samples:

$$\sigma_{\bar{x}_1 - \bar{x}_2} = \sigma \sqrt{\frac{1}{n_1'} + \frac{1}{n_2'}} \qquad (8.78)$$

and this should be used in the construction of confidence intervals for the difference in means. Now the ratio of the two variances is

$$\psi^{(2)} = \frac{\left(\dfrac{1}{n_1} + \dfrac{1}{n_2} \right)}{\left(\dfrac{1}{n_1'} + \dfrac{1}{n_2'} \right)} = \frac{(n_1 + n_2)(\psi_1 \psi_2)}{n_1' + n_2'} \le 1 \qquad (8.79)$$

where ψ_1 and ψ_2 are defined as before, and the superscript "(2)" is used to denote the ratio of the variances for the two-sample case.

For the CAR model,

$$\psi^{(2)} = \frac{\psi_1 \psi_2 (n_1 + n_2)}{n_1 + n_2 - \rho_1 \Sigma_{11} + \rho_2 \Sigma_{12}} \qquad (8.80)$$

where there are possibly different degrees of spatial dependence in each sample, and where Σ_{1j} refers to the sum of the weight matrix for sample j. Also, the values of ψ_1 and

ψ_2 are found from (8.56). In the special case of row-standardized weight matrices, and where $\rho_1 = \rho_2 = \rho$,

$$\psi^{(2)} = 1 - \rho \tag{8.81}$$

For the SAR model,

$$\psi^{(2)} = \frac{\psi_1 \psi_2 (n_1 + n_2)}{n_1 + n_2 - \rho_1 [2\Sigma_{11} + \rho_1 \Sigma_{21}] + \rho_2 [2\Sigma_{12} + \rho_2 \Sigma_{22}]} \tag{8.82}$$

where Σ_{2j} refers to the sum of the $\mathbf{W}^2$ matrix for sample j and where the values of ψ_1 and ψ_2 are found from (8.53). In the special case of row-standardized matrices and where $\rho_1 = \rho_2 = \rho$,

$$\psi^{(2)} = (1 - \rho)^2 \tag{8.83}$$

These results can then be used to find the probability of Type I and Type II errors, and corrected critical values and p-values for two-sample tests as follows:

$$\alpha' = 1 - \Phi\left(1.645\sqrt{\psi^{(2)}}\right) \tag{8.84}$$

$$\beta' = \Phi\left(\frac{1.645}{\sqrt{\psi^{(2)}}} - \frac{\delta\sqrt{n'}}{\sigma}\right) \tag{8.85}$$

$$z'_{crit} = \frac{\Phi^{-1}(1-\alpha)}{\sqrt{\psi^{(2)}}} = z_{crit} / \sqrt{\psi^{(2)}} \tag{8.86}$$

$$p' = p / \sqrt{\psi^2} \tag{8.87}$$

The preceding effects and adjustments assume that ρ and σ are known. When ρ and σ are unknown and must be estimated, ρ may first be estimated using Equation 8.34 when the data have been detrended and E[z]=0, and Equation 8.37 when the mean is unknown. The result can be used to find the maximum likelihood estimate of the unknown mean (Equation 8.9), and the variance can be estimated from

$$\hat{\sigma}^2 = \left(\mathbf{x} - \tilde{\mu}\right)' \mathbf{V}^{-1} \left(\mathbf{x} - \tilde{\mu}\right) \tag{8.88}$$

It is then straightforward to use these estimates of the mean and variance to carry out tests of inference. It is also possible, using the estimates of ρ and σ, and the results above, to estimate the effects on p-values, critical values, and Type I and Type II errors, *if the* dependence is ignored.

8.4.5 Illustrations

Testing Null Hypotheses About the Mean

Suppose that we were interested in testing the null hypothesis that the true mean is equal to zero. With a two-sided test with $\alpha = 0.05$ and a large sample size, the usual critical values of z are ± 1.96. With spatial dependence, too many null hypotheses will be rejected if these values are used.

Assume first that σ is known, and equal to one. As an example an SAR process was simulated on a 5 × 5 grid with rook's adjacency, with $\mu = 0$, $\sigma = 1$, and $\rho = 0.6$. The standard z-test of the null hypothesis that the true mean was zero, using $\alpha = 0.05$ and $z = (\bar{x} - 0)/(1/\sqrt{n})$, was rejected in 42.9% of the 10,000 simulations. The 2.5th and 97.5th percentiles of the simulated distribution were −4.92 and 4.97. This over-rejection of true null hypotheses can be addressed by either (a) adjusting the test statistic to account for the dependence, or (b) adjusting the critical values of the test statistic.

The statistic can be adjusted by using the effective sample size (n') in place of the actual sample size. Thus $z = (\bar{x} - 0)/(1/\sqrt{n'})$, where $n' = 1'V^{-1}1$. Note that this adjusted statistic is equal to the usual z-statistic, multiplied by $\sqrt{\psi}$, where ψ is the ratio of effective sample size to actual sample size ($n'/n = 1' V^{-1}1/n$). In this example $n' = 4$, and when this is used, the critical value was exceeded in 5.29% of the 10,000 simulations, close to the nominal value of 5%. The 2.5th and 97.5th percentiles of the simulated z-statistics were −1.96 and 1.99, respectively.

An estimate of the Type I error rate for two-tailed tests, when dependence is ignored, can be found from twice the tail area, $2(1 - \Phi(1.96 \sqrt{\psi}))$. In this example, $\psi = 4/25$, and the estimate of the Type I error rate is $2(0.2165) = 0.433$, close to the fraction of null hypotheses rejected in the simulation described above (0.429).

But the previous example is a bit unrealistic, since we rarely if ever know the true variance, σ^2. It is more common that we will have to estimate μ, σ, and ρ. Usually, with 25 observations, we would use a t-test with 24 degrees of freedom. For a two-sided test with $\alpha = 0.05$, the critical values are ± 2.064. The test statistic is

$$t = (\bar{x} - 0)/(s/\sqrt{n}) \tag{8.89}$$

For 10,000 simulations of the spatially dependent process described above, the null hypothesis was rejected 37.1% of the time, and the 2.5th and 97.5th percentiles of the simulated distribution of t-statistics were −4.56 and 4.57, respectively.

To adjust the t-statistic, we need to again use n' instead of n, and we also need to correct for the fact that the variance of the data is inflated in the presence of spatial dependence. Haining (1988), for example, gave

$$E\left[s^2\right] = \frac{1}{n-1}\left(\text{Tr}[\mathbf{V}] - \frac{\mathbf{1'V1}}{n}\right)$$ (8.90)

as the estimated value for the true variance σ^2, in the presence of spatial autocorrelation. In the present example,

$$E\left[s^2\right] = \frac{1}{24}\left(39.4 - \frac{157.9}{25}\right) = 1.379$$ (8.91)

This was verified by again simulating the process 10,000 times; the mean of the 10,000 sample variances was 1.3779.

An estimate of the Type I error in this scenario can be found by reducing the estimate in the previous example (where σ is assumed known) by a factor of $\sqrt{E\left[s^2\right]} = \sqrt{1.379}$. Thus we would expect a Type I error level of $0.433/\sqrt{1.379} = 0.369$ of the time; the simulated error rate of 37.1% reported above is very close to this.

How should we adjust our statistic? We use n', and we use the maximum likelihood estimate for σ^2 (Equation 8.13), which accounts for the dependence. Alternatively, we can multiply our usual t-statistic by s/σ_{ML}, or we multiply our usual t-statistic by $s/\sqrt{E\left[s^2\right]}$. Thus, when we use

$$t = \frac{\bar{x} - 0}{\left(\dfrac{\hat{\sigma}_{ML}}{\sqrt{n'}}\right)}$$ (8.92)

with the same critical values (±2.064), the null hypothesis is rejected 5.3% of the time in 10,000 simulations. The 2.5th and 97.5th percentiles of the distribution of simulated statistics were −2.056 and 2.108, close to the actual critical value of 2.064.

As an alternative to using adjusted test statistics, we could use adjusted critical values. Instead of using critical values of ±1.96 in the z-test above we should instead use values of $\pm 1.96/\sqrt{\psi}$. In our example with $\psi = 0.16$, this implies critical values of ±4.9. Note first that these values are close to the simulated 2.5th and 97.5th percentiles in the first simulation, where σ is assumed known. Ten thousand simulations of the z-test with critical values of ±4.9 yielded a Type I error level of 0.0505.

When σ is unknown, we can adjust the critical values using knowledge of both the effective sample size n' and the effect of dependence on the estimate of the variance. The critical value of 2.064 used in the t-test above can be adjusted by dividing by $\sqrt{\psi}$ to account for the effective sample size, and then dividing the result by the square root of $E[s^2]$ to account for the effects of dependence on the variance. Here this leads to an adjusted critical value of $[2.064/(0.4)]/\sqrt{1.379} = 4.40$, which is not too far from the simulated 2.5th and 97.5th percentiles of −4.56 and 4.57, reported above. The small difference between the approximate and simulated critical values is due to the relatively small sample size.

Summary

For the SAR model, the adjustments to Type I error calculations and critical values can be made easily, since

$$n' = n - \rho\left(2\Sigma_1 - \rho\Sigma_2\right)$$ (8.93)

where $\Sigma_1 = \mathbf{1'\,W1}$ is the sum of the elements of $\mathbf{W}$, and Σ_2 is the sum of the elements of $\mathbf{W}^2$. Thus the quantity ψ is equal to

$$\psi = \frac{n'}{n} = 1 - \left(\frac{\rho}{n}\right)\left(2\Sigma_1 - \rho\Sigma_2\right)$$ (8.94)

This leaves the analyst with several alternatives:

(a) Use the usual z-test for hypotheses about the mean, but adjust the critical values by dividing the usual value by $\sqrt{\psi}\,\sqrt{E\left[s^2\right]/s^2}$.

(b) Adjust the test statistic by multiplying the usual statistic by $\sqrt{\psi}\,\sqrt{s^2/E\left[s^2\right]}$.

(c) Use the maximum likelihood estimate of the mean and its associated standard error to test hypotheses about the mean.

In addition, estimates of the Type I error that would result from proceeding with hypothesis tests that did not have corrections for spatial dependence can be found by finding tail areas associated with $z_{crit}\sqrt{\psi}$.

Mercer and Hall Wheat Yield Data

Mercer and Hall (1911) reported on the wheat yields for a one acre plot at Rothamsted, England in 1910. Data are for cells from a 20×25 lattice, where the rows go from west to east, and where the columns go from north to south. The mean yield across the $n = 500$ cells is 3.949; the standard deviation is equal to 0.458. A standardized weight matrix is defined by first using binary adjacency, and then dividing each row by its sum. Observations were standardized by first subtracting the mean, and then dividing the result by the standard deviation.

Using the approximate profile likelihood estimator of Li et al. (2007) to estimate the spatial dependence parameter when fitting a SAR model, $\hat{\rho} = 0.574$ and this is very close to the value of $\hat{\rho} = 0.573$ arrived at by using the alternative estimator suggested by Ord (1975) as the solution to the following equation:

$$\mathbf{z'}(\mathbf{I} - \rho\mathbf{W})'\mathbf{W}(\mathbf{I} - \rho\mathbf{W})\mathbf{z} = 0$$ (8.95)

Now

$$\psi = \frac{n'}{n} = (1-\rho)^2 = 0.181$$ (8.96)

implying that the sample size of 500 is, due to the spatial dependence, equivalent to $n' = 500(0.181) = 90.5$ effectively independent observations. In the absence of any adjustment, a one-sided test of the mean would have a Type I error probability of about 24%, instead of the usual 5%:

$$\alpha' = 1 - \Phi[1.645(1 - \rho)] = 0.242 \tag{8.97}$$

$$\alpha' \approx \frac{1 - \sqrt{1 - e^{-\frac{2(z_{crit}(1-\rho))^2}{\pi}}}}{2} = 0.241 \tag{8.98}$$

A corrected critical value for a one-tailed alternative, with $\alpha = 0.05$, is $1.645/(1 - 0.574) = 3.862$; this would yield the desired Type I error probability of 0.05. We of course do not know the "true" mean (μ_0) in this example, but if the null hypothesis, whatever the value of μ_0, were true, this relatively high amount of spatial dependency would cause samples of 500 to reject that true null hypothesis almost a quarter of the time.

When binary, rook's adjacency is used to define the weight matrix without standardization, $\Sigma_1 = 1910, \Sigma_2 = 7378, \hat{\rho}_N = 0.153$ and

$$\psi = \frac{n'}{n} = 1 - \left(\frac{\rho}{n}\right)(2\Sigma_1 - \rho\Sigma_2) = 0.177 \tag{8.99}$$

$$\alpha' = 1 - \Phi\left(1.645\sqrt{\psi}\right) \approx \frac{1 - \sqrt{1 - e^{-\frac{2(1.645\sqrt{\psi})^2}{\pi}}}}{2} = 0.244 \tag{8.100}$$

with these latter values close to the values found using the standardized weight matrix, as expected.

To illustrate the application of a two-sample test, suppose that we wish to test the null hypothesis that the mean in the northern 10×25 portion of the grid is no different from the mean in the southern 10×25 portion. The sample mean and standard deviation are 3.965 and 0.446 for the northern half of the study area; the corresponding values for the southern half are 3.9833 and 0.471. The usual difference of means test yields

$$z = \frac{3.965 - 3.983}{0.458\sqrt{\frac{1}{250} + \frac{1}{250}}} = -0.31 \tag{8.101}$$

and since this is less than the critical value of z, there is not enough evidence to reject the null hypothesis. Considering spatial dependency in the observations, the approximate profile likelihood estimator for the SAR model yields $\hat{\rho}_N = 0.497$ for the northern half and $\hat{\rho}_S = 0.602$ for the southern half. Therefore, $\psi_N = (1 - 0.497)^2 = 0.253$ and $\psi_S = (1 - 0.602)^2 = 0.158$. This in turn implies effective sample sizes of $n'_N = (0.253) \times 250 = 63.25$ and $n'_S = 0.158 \times (250) = 39.5$, in the north and south, respectively. The ratio of the variances of the distribution of differences in sample means (not considering dependency, relatively to considering dependency), using Equation 8.79, is $\psi^{(2)} = 0.195$. Using (8.84), this implies a Type I error probability of 0.234 if results are uncorrected; the corrected critical value for a one-tailed test with $\alpha = 0.05$ would be $1.645/\sqrt{0.195} = 3.73$.

Summary

With simultaneous autoregressive and conditional autoregressive spatial models, the inverse of the variance–covariance matrix describing the relationships between pairs of observations has a simple form. The sum of the elements of this matrix is easy to obtain and it represents the effective sample size n', accounting for spatial dependence. This makes it straightforward to (a) find corrected critical values and p-values for statistical tests, and (b) find the probability of Type I and Type II errors, accounting for spatial dependence. The results of Li et al. (2007, 2012) simplified the estimation of the spatial dependence parameter for the SAR model. Coupled with the results here, this should facilitate the calculation and reporting of results from statistical analyses of spatial data.

8.5 Spatial Regression

The spatial errors model (SEM) in a regression context may be stated as

$$Y = X\beta + \epsilon \tag{8.102}$$

where ϵ is normally distributed with mean 0 and variance–covariance matrix $V = \{(I-\rho W)'(I-\rho W)\}^{-1}$. The matrix X has dimension $n \times (p + 1)$, where there are n observations. The first column of the matrix consists of a set of ones, and each of the other columns corresponds to one of the p independent, explanatory variables. The errors are assumed to follow a spatial, autoregressive process:

$$\epsilon = \rho W \epsilon + v \tag{8.103}$$

where ρ and W are as defined previously, and v is an $n \times 1$ vector of independent, normally distributed error terms. Thus the errors from an ordinary least squares solution of (8.102) are spatially autocorrelated and the regression in (8.102) can be written as

$$Y = X\beta + (I-\rho W)^{-1} v \tag{8.104}$$

Alternatively, this can be expressed as

$$(I-\rho W)Y = (I-\rho W)X\beta + v \tag{8.105}$$

Since the elements of v are independent, if ρ is known, ordinary least squares can be carried out with this equation, regressing the transformed variables $\mathbf{Y}^* = (\mathbf{I} - \rho \mathbf{W})\mathbf{Y}$ on the transformed regresssors $\mathbf{X}^* = (\mathbf{I} - \rho \mathbf{W})\mathbf{X}$. However, it is of course not usually (if ever!) known.

In addition to the regression coefficients (β), we need to estimate the variance of the errors (σ^2) and the spatial dependence parameter, ρ. A starting point for estimation is the log likelihood of the data, given the model

$$\ln \mathcal{L} = \ln |\mathbf{V}^{-1}| + n \, \ln(\mathbf{Y} - \mathbf{X}\beta_{GLS})'(\mathbf{Y} - \mathbf{X}\beta_{GLS}) \tag{8.106}$$

where β_{GLS} is the estimate of the regression coefficients found from generalized least squares:

$$\beta_{GLS} = (\mathbf{X}'\mathbf{V}^{-1}\mathbf{X})^{-1}\mathbf{X}'\mathbf{V}^{-1}\mathbf{Y} \tag{8.107}$$

Equation 8.106 may be rewritten as

$$\ln \mathcal{L} = \ln |\mathbf{V}^{-1}| + n \ln(\mathbf{R}\mathbf{Y})' \mathbf{V}^{-1}(\mathbf{R}\mathbf{Y}) \tag{8.108}$$

where $|\mathbf{V}^{-1}|$ refers to the *determinant* of $\mathbf{V}^{-1}$, and where

$$\mathbf{R} = \mathbf{I} - \mathbf{X}'(\mathbf{X}\mathbf{V}^{-1}\mathbf{X})^{-1}\mathbf{X}'\mathbf{V}^{-1} \tag{8.109}$$

This is known as a profile or condensed likelihood function, where the maximum likelihood estimates of the other parameters have been substituted into the original likelihood function (8.106), in order to have a function that is solely a function of one parameter (in this case, ρ).

So, Equation 8.108 is now a function only of the parameter ρ. The first step is to maximize Equation 8.108 with respect to ρ, by numerical methods, or by a brute-force search using different values of ρ (since it is not possible to derive a simple analytical expression for ρ by setting the derivative of (8.108) equal to 0 and solving for ρ). Then the maximum likelihood estimates of β and σ^2 may be found from

$$\beta = (\mathbf{X}'\mathbf{V}^{-1}\mathbf{X})^{-1}\mathbf{X}'\mathbf{V}^{-1}\mathbf{Y} \tag{8.110}$$

and

$$\sigma^2 = \frac{1}{n}(\mathbf{Y} - \mathbf{X}\beta)'\mathbf{V}^{-1}(\mathbf{Y} - \mathbf{X}\beta) \tag{8.111}$$

where, as a reminder, $\mathbf{V}^{-1} = (\mathbf{I} - \rho \, \mathbf{W}')(\mathbf{I} - \rho \, \mathbf{W})$ is used with the estimate of ρ.

As an alternative to using numerical methods or a brute-force search to estimate ρ, Li et al. (2012) approximated (8.108) with a linear equation and then found an estimate.

In this case, the variance–covariance matrix of the estimates of the regression coefficients is $\sigma^2 (\mathbf{X}'\mathbf{V}^{-1}\mathbf{X})^{-1}$, and hence the ratio of the variance of the estimate when accounting for dependence, to the variance of the estimate when not accounting for it, is $\hat{\sigma}^2 \left(\mathbf{X}'\mathbf{V}^{-1}\mathbf{X} \right)^{-1} / \hat{\sigma}_e^2 \left(\mathbf{X}'\mathbf{X} \right)^{-1}$, where $\hat{\sigma}_e^2$ is the estimate of the variance of the residuals using ordinary least squares and ignoring spatial dependence.

Example

A hypothetical five-region system was constructed with the following row-standardized weight matrix:

$$\mathbf{W} = \begin{bmatrix} 0 & 1 & 0 & 0 & 0 \\ \frac{1}{2} & 0 & \frac{1}{2} & 0 & 0 \\ 0 & \frac{1}{3} & 0 & \frac{1}{3} & \frac{1}{3} \\ 0 & 0 & \frac{1}{2} & 0 & \frac{1}{2} \\ 0 & 0 & \frac{1}{2} & \frac{1}{2} & 0 \end{bmatrix} \tag{8.112}$$

Data were generated from a "true" linear relationship

$$y = 3 + 10x_1 + e \tag{8.113}$$

The errors were spatially autocorrelated and generated from $(\mathbf{I} - \rho\mathbf{W})^{-1}v$, where v was a vector of five random independent variables from a normal distribution with mean 0 and standard deviation equal to 6. This resulted in the following "observed" data:

$$\mathbf{X} = \begin{bmatrix} 1 \\ 2 \\ 3 \\ 4 \\ 9 \end{bmatrix}; \mathbf{Y} = \begin{bmatrix} 11.95 \\ 21.23 \\ 34.18 \\ 45.09 \\ 93.67 \end{bmatrix} \tag{8.114}$$

The ordinary least squares estimate of the coefficients is

$$\beta = \begin{bmatrix} 2.334 \\ 10.234 \end{bmatrix} \tag{8.115}$$

and the sum of squared residuals was $\mathbf{e}'\mathbf{e} = 8.067$. Maximization of Equation (8.106) or (8.108) leads to $\check{\rho}_{ML} = 0.1516$. The APLE estimator (Equation 8.37) is $\check{\rho}_{APLE} = 0.153$. The regression estimates associated with these are, from Equation 8.110,

$$\beta_{ML} = \begin{bmatrix} 2.695 \\ 10.137 \end{bmatrix}; \beta_{APLE} = \begin{bmatrix} 2.692 \\ 10.138 \end{bmatrix} \tag{8.116}$$

and the estimates of the error variances from Equation 8.111 are 8.041/5 and 8.043/5 for the ML and APLE estimators, respectively.

Exercise

1. There are four values observed within a region: $x_1 = 0.34$; $x_2 = 0.77$; $x_3 = 1.12$; $x_4 = 2.00$.
   Assume independent observations:

   (a) Find the simple average.
   (b) Assume the true standard deviation is known, and is equal to 1. What is the standard deviation of the mean?
       Now assume dependent observations.
       For the elements of the variance–covariance matrix **V**, the diagonals are equal to one, since this is the known variance. Also let $v_{23} = v_{32} = 0.7$; $v_{12} = v_{21} = 0.2$; $v_{13} = v_{31} = 0.1$. Assume all other elements of the matrix are equal to zero.
   (c) Find **1'V1**, the sum of the elements of **V**. Use this to find the standard deviation of the mean. Your answer should be greater than that found in (a) above.
   (d) Find **V⁻¹**.
   (e) Find the column sums of **V⁻¹** and use them as weights to find the maximum likelihood estimate of the mean (remember to divide the weighted sum of observations by the sum of the weights).
   (f) Find **1'V⁻¹1**, which is the effective sample size. (This is also the sum of the weights you found in (e).) Find its reciprocal, which is the maximum likelihood estimate of the variance of the sample mean (which should be smaller than the answer in (c)).

Further reading

The paper by Haining (1988) is one of the few that sets out the elementary questions (and solutions) associated with the simple question of finding the mean for a set of spatial data. There are a large number of books that cover the topic of spatial regression. The book by Bailey and Gatrell (1995) has a nice introduction that is written at an accessible level. It is impossible to choose from the many other introductions; however, one is Anselin (2009).

Online resources

Visit **https://study.sagepub.com/rogersonspatialstatistics** for resources which supplement the material presented in this chapter. Students can find datasets, additional exercises and sample code files, which can be used to practice the techniques covered in this chapter, as well as a chapter introduction video and video demonstration of implementing spatial regression techniques in GeoDa. The topic-software chart also describes the software packages that can be used to implement key techniques.

Lecturers and instructors can find exercises and conceptual prompts to facilitate classroom discussion and practice.

Epilogue

This book by no means serves as a comprehensive introduction to the field of spatial statistics. Point process models as well as the concepts and methods associated with the subfield of geostatistics (e.g., kriging and variograms) are two examples of topics that are not covered here. Space–time models are not covered at all, and these are an area of active research. Spatial regression methods and models have been set out in only an elementary way in Chapter 8 and there is a wealth of material – in the form of both books and research articles – that may be accessed for more in-depth coverage of all of these topics. Some additional resources that cover these and other topics often treated in more advanced and specialized texts are listed below.

Schabenberger, O. and Gotway, C.A. (2005) *Statistical Methods for Spatial Data Analysis*. Boca Raton, FL: Chapman & Hall/CRC.

Covers autocorrelation, semivariograms, random fields, mapped point patterns all in the first three chapters, and then goes more deeply into these topics and spatial regression in subsequent chapters.

Gelfand, A.E., Diggle, P.J., Fuentes, M. and Guttorp, P. (2010) *Handbook of Spatial Statistics*. Boca Raton, FL: Chapman & Hall/CRC.

This is a set of 31 contributions containing sections on continuous and discrete variation, point patterns, and space–time processes.

Chun, Y. and Griffith, D.A. (2013) *Spatial Statistics and Geostatistics: Theory and Applications for Geographic Information Science and Technology*. London: Sage.

Covers spatial autocorrelation, spatial sampling, spatial interpolation, spatially adjusted regression, and local statistics. R code is included with each chapter.

LeSage, J. and Pace, R.K. (2009) *Introduction to Spatial Econometrics*. Boca Raton, FL: Chapman and Hall/CRC.

This book is, as the name implies, heavily focused on spatial models and spatial econometrics.

Anselin, L. (1988) *Spatial Econometrics: Methods and Models*. Dordrecht: Kluwer Academic.

An early but classic treatment of spatial econometrics.

Cressie, N.A.C. (2015) *Statistics for Spatial Data* (revised edition). New York, NY: Wiley.

This is a comprehensive and fairly advanced treatment of spatial statistics from a statistician's perspective. This is also perhaps the most popular and widely accepted treatment of the topics in this field.

Cressie, N.A.C. and Wikle, C.K. (2011) *Statistics for Spatio-Temporal Data*. Hoboken, NJ: Wiley.

An award-winning recent treatment of space–time statistics.

Bivand, R.S., Pebesma, E., and Gomez-Rubio, V. (2013) *Applied Spatial Data Analysis with R* (2nd edition). New York, NY: Springer.

Several chapters in the beginning of the book are devoted to the handling of spatial data. Attention is next given to many methods of spatial analysis, spatial statistics, and geostatistics, including several of the methods described in this book (e.g., Kulldorff's spatial scan statistic, Moran's *I*, and Stone's test.)

Fischer, M.M. and Getis, A. (2009) *Handbook of Applied Spatial Analysis: Software Tools, Methods, and Applications*. Berlin: Springer.

This edited collection contains a wide range of contributions focusing on software, many of the major topics in the fields of statistics and geostatistics, and applications in the areas of remote sensing, economics, and the analysis of the environment and health.

Waller, L.A. and Gotway, C.A. (2004) *Applied Spatial Statistics for Public Health Data*. New York, NY: Wiley.

A student-oriented book written at a level similar to this book, with applications to health.

Brunsdon, C. and Comber, L. (2015) *An Introduction to R for Spatial Analysis and Mapping*. London: Sage.

This is a bit similar to the book by Bivand et al., but a bit more recent and perhaps written for a slightly less advanced audience. This book covers the ins and outs of both R and the handling of spatial data. There is good coverage of using R as a GIS and using R with the internet. Also explained is the use of R with the basic methods of point pattern analysis, Moran's *I*, and spatial regression.

APPENDIX A

Some Preparatory Tools

In this appendix, we extend elementary concepts of probability and statistics, and develop some of the tools that are useful in the study of spatial patterns and spatial statistical analysis. These include:

- calculus (Section A.1),
- matrix algebra (Section A.2), including application to regression analysis,
- review and extension of some probability theory (Section A.3),
- parameter estimation (Section A.4), and
- simulation of variates from probability distributions (Section A.5).

A.1 A Calculus Primer: Derivatives and Integrals

In a few sections of the book we use the tools of calculus. We use only the most elementary concepts, and therefore the level of technical detail and explanation should not be unduly challenging.

There are two primary areas of application that we explore. One is the use of integrals and the tools of integration to find areas under curves or functions (probabilities correspond to areas under the curves – known as probability density functions or pdfs – describing continuous random variables). The second area of application is to use derivatives to find the slope of a curve or function; by setting the slope equal to zero, we can find the point at which the function is maximized or minimized. This will be useful, for example, in the estimation of parameters describing the shape of probability density functions.

In this section, we provide an introduction/review of these important concepts and tools from calculus.

A.1.1 Integrals as Areas Under Curves

Integrals can be thought of both as infinite sums and as areas under curves. For the histograms associated with discrete random variables, we are often interested in the sum of probabilities across several possible outcomes; this corresponds to summing the heights of the vertical bars of the histogram which has relative frequency on the vertical axis. For continuous random variables, the idea is the same, but we have an infinite number of infinitely narrow vertical bars – and the summation sign is replaced with an integral sign. Seeing an integral sign should not cause undue stress – simply think of it as a summation sign, where there are an infinite number of items to sum.

Consider the uniform random variable; its density is given by

$$y = f(x) = \frac{1}{b-a}; a \le x \le b \tag{A.1}$$

This is a curve that traces out a straight horizontal line at a height of $y = 1/(b-a)$, over the range from $x = a$ to $x = b$. Of course the area of the full rectangle is 1, since the probability of getting some value in the range between a and b is equal to 1. The rectangle has a length of $b - a$, and a height of $1/(b-a)$. The probability that this variable takes on a value between c and d is equal to the area of the shaded rectangle in Figure A.1. This is equal to $\{1/(b-a)\}\{d-c\} = (d-c)/(b-a)$.

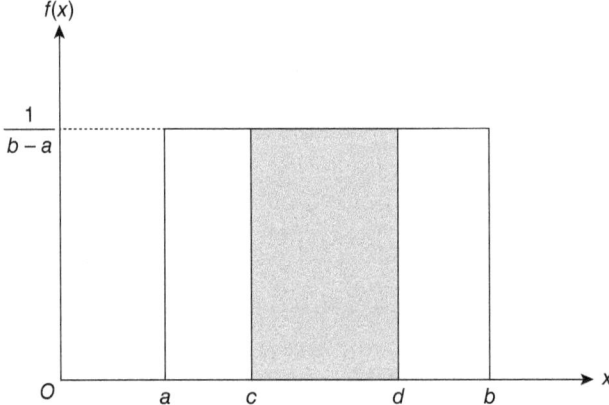

Figure A.1   Uniform probability distribution

We can think of this area as the sum of the areas of many (in fact, an infinity of) very narrow vertical bars (after all, a continuous random variable is simply a histogram, with an infinite number of vertical bars), all with height $1/(b-a)$. Let dx be the very small width of one of these bars. Then we can express the area as

$$\int_c^d \frac{1}{b-a} dx \tag{A.2}$$

where the integral sign ($\int$) is used instead of a summation sign, and the beginning and end points of the infinite summation are placed at the bottom and top of the integral sign, respectively (paralleling what is done with summation signs). To evaluate this integral, we first use the rule that the integral of a constant is equal to that constant, times x. That is, $\int k\,dx = kx$, where k is a constant (this is termed an "indefinite integral"; when there are limits on the integral sign in a specific example, it is termed a "definite integral"). We then evaluate this result by (a) substituting the upper limit for x in the result, and then (b) subtracting from the result in (a) the result when the lower limit is substituted for x. Thus

$$\int_c^d \frac{1}{b-a}\,dx = \frac{x}{b-a}\Big|_c^d = \frac{d}{b-a} - \frac{c}{b-a} = \frac{d-c}{b-a} \tag{A.3}$$

The area under other simple curves is also easy to find geometrically. For example, if $f(x) = 3x$, the area under the function between $x = 2$ and $x = 4$ is the area of the trapezoid in Figure A.2.

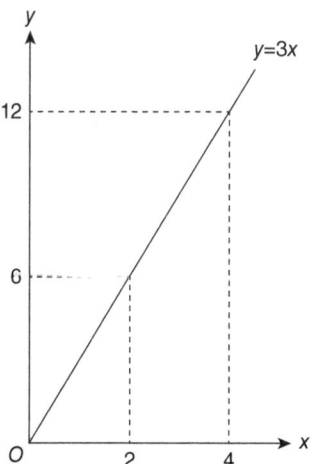

Figure A.2 Integrals as areas under curves

Recall that the area of a trapezoid is equal to one-half the sum of the parallel bases, times the height. Here we have $(6 + 12)/2$, times the height of $(4 - 2)$; the area is equal to $(9) \times (2) = 18$.
We can also use the fact that $\int cx\,dx = cx^2/2$ to find

$$\int_2^4 3x\,dx = 3\int_2^4 x\,dx = \frac{3x^2}{2}\Big|_2^4 = \frac{3(4^2)}{2} - \frac{3(2^2)}{2} = \frac{48}{2} - \frac{12}{2} = 18 \tag{A.4}$$

Table A.1 provides a short table of some common definite integrals.

Table A.1   Common indefinite integrals

Function	Indefinite integral
$f(x)$	$\int f(x)$
x	$x^2/2$
x^n	$x^{n+1}/(n+1)$
e^x	e^x
$1/x$	$\ln\|x\|$
$\cos(x)$	$\sin(x)$
$\sin(x)$	$-\cos(x)$

A.1.2 Areas Under Probability Density Functions as Probabilities

Continuous probability distributions have the property that the area under the distribution sums to 1. If for example a random variable took on x values between 1 and 3, and had the form $f(x) = c - 0.5x$ (where c is a constant), we could solve

$$\int_1^3 (c - 0.5x)\, dx = 1 \tag{A.5}$$

to find the value of c that would make this a valid probability density function. Thus

$$\int_1^3 (c - 0.5x)\, dx = \int_1^3 c\, dx - \int_1^3 0.5x\, dx = 1 \tag{A.6}$$

and this implies that the definite integral is equal to

$$cx\Big|_1^3 - \frac{0.5x^2}{2}\Big|_1^3 = (3c - c) - \left(\frac{3^2}{4} - \frac{1}{4}\right) = 2c - 2 = 1 \tag{A.7}$$

and therefore $c = 1.5$.

A.1.3 Derivatives

Derivatives of functions are themselves functions that, at a given value x, are equal to the slope of the tangent to the original function at x. If the original function is denoted $f(x)$, the derivative is denoted $\frac{d}{dx} f(x) = f'(x)$. For example, the derivative of a straight line function is equal to a constant, which in turn is equal to the slope of that line. Thus for the function $y = f(x) = a + bx$, the derivative is equal to $f'(x) = b$. Table A.2 provides a short list of common derivatives.

Table A.2 *Common derivatives*

f(x)	f'(x)
e^x	e^x
a	0
ax	a
ax^b	abx^{b-1}
$\ln(x)$	$1/x$
a^x	$a^x \ln(a)$
$\sin(x)$	$\cos(x)$
$\cos(x)$	$-\sin(x)$

In the table, a and b are constants, and x is a variable.

One important rule associated with differentiation is the *chain rule*. This can be stated as follows, letting $F(x) = f(g(x))$:

$$F(x) = f'(g(x))g'(x) \tag{A.8}$$

For example, suppose that

$$F(x) = e^{2x} \tag{A.9}$$

We know that the derivative of e to a power is equal to e to that power. But in this case that power itself has a derivative ($g(x) = 2x$), and the derivative of that power is also included, as a product, in the derivative:

$$F'(x) = e^{2x}(2) = 2e^{2x} \tag{A.10}$$

To take another example, suppose that $F(x) = \sqrt{x^3 - 2x}$. This can also be written as $F(x) = (x^3 - 2x)^{1/2}$.

Using the table of derivatives, we have $F'(x) = (1/2)(x^3 - 2x)^{-1/2}$, but the chain rule tells us that we also need to take the derivative of what is inside the parentheses. In this case $g(x) = x^3 - 2x$ and $g'(x) = 3x^2 - 2$. Therefore, the final result is $F'(x) = (1/2)(3x^2 - 2)(x^3 - 2x)^{-1/2}$.

Note the relationship between integrals and derivatives. The derivative of an integral of a function is that function. Similarly, the integral of a derivative of a function is that function (plus a constant). A function $g(x)$ can be thought of as the slope for another function, $f(x)$. When we know $g(x)$, we can use integrals to find $f(x)$.

We'll use derivatives in maximizing functions by finding the point where the slope of the tangent to the function is equal to zero. One such application is where we will want to maximize the likelihood of our observations to estimate the parameters of models and distributions (see Section A.4.3).

A.2 Matrix Algebra: A Short and Gentle Introduction

It is often convenient to arrange information compactly into tables, or matrices. These matrices can be manipulated using rules of matrix algebra which in many cases are very similar to the more well-known rules of algebra. By expressing information and models in matrix form, it becomes easier to deepen understanding and allow generalizations of familiar statistical methods. An important example is the expression of the linear regression model in matrix form. By writing the linear regression model in matrix form, it is relatively straightforward to understand extensions to, for example, the study of outliers, the treatment of deviations from assumptions (such as homoscedasticity and independence of disturbances), and the treatment of multicollinearity (i.e., correlation) among the independent variables.

A matrix is said to have order $m \times n$ when it has m rows and n columns. The entries of a matrix A are denoted $\{a_{ij}\}$, where the subscript i refers to the row, and the subscript j refers to the column. A matrix with one column is a *column vector*; a matrix with one row is termed a *row vector*. A matrix with just one row *and* one column is a *scalar*. Matrices are typically notated with bold, upper case letters, and the individual elements of the matrix are notated with lower case letters. Thus

$$A_{2x3} = \begin{bmatrix} a_{11} & a_{12} & a_{13} \\ a_{21} & a_{22} & a_{23} \end{bmatrix} \tag{A.11}$$

Matrices of the same order may be added and subtracted, and these operations proceed element by element. Thus if $C = A + B$, the elements of C are found from $c_{ij} = a_{ij} + b_{ij}$.

Example:

$$A = \begin{bmatrix} 4 & 6 & 11 & 21 \\ 5 & 8 & 32 & 41 \end{bmatrix}; B = \begin{bmatrix} 10 & 12 & 14 & 16 \\ 32 & 55 & 11 & 10 \end{bmatrix} \tag{A.12}$$

$$A + B = \begin{bmatrix} 14 & 18 & 25 & 37 \\ 37 & 63 & 43 & 51 \end{bmatrix} \tag{A.13}$$

Subtraction is also carried out element by element:

$$B - A = \begin{bmatrix} 6 & 6 & 3 & -5 \\ 27 & 47 & -21 & -31 \end{bmatrix} \tag{A.14}$$

A.2.1 Matrix Multiplication

A prerequisite for post-multiplying a matrix A by a matrix B to form the product $C = AB$ is that the number of columns in A must be equal to the number of rows in B. The result is a matrix, say C, which has number of rows equal to the number of rows in A, and number of columns equal to the number of columns in B.

Each element in the product matrix, c_{ij}, is equal to the following sum of products:

$$c_{ij} = \sum_{k=1}^{n} a_{ik} b_{kj} \tag{A.15}$$

where n is the number of columns in A and the number of rows in B. For example, if

$$A = \begin{bmatrix} 2 & 6 & 3 \\ 4 & 7 & 2 \end{bmatrix}; \ B = \begin{bmatrix} 8 & 2 \\ 5 & 9 \\ 4 & 1 \end{bmatrix} \tag{A.16}$$

we can find the product AB since the number of rows in A is equal to the number of columns in B. Furthermore, the product matrix $C = AB$ has order 2×2. To illustrate,

$$c_{11} = 2(8) + 6(5) + 3(4) = 58 \tag{A.17}$$

Similarly,

$$c_{12} = 2(2) + 6(9) + 3(1) = 61$$

$$c_{21} = 4(8) + 7(5) + 2(4) = 75 \tag{A.18}$$

$$c_{22} = 4(2) + 7(9) + 2(1) = 73$$

Therefore, in this example,

$$C = \begin{bmatrix} 58 & 61 \\ 75 & 73 \end{bmatrix} \tag{A.19}$$

If A is $m \times n$ and B is $n \times p$, then AB is of order $m \times p$. Note that matrix multiplication is not commutative; AB is not generally equal to BA. It is, however, associative and distributive: $(AB)C = A(BC)$ and $A(B + C) = AB + AC$.

A.2.2 Other Terminology and Properties

The *transpose* of a matrix A is denoted by A' and is found by interchanging the rows and columns; the first row becomes the first column of the transpose, the second row becomes the second column of the transpose, etc. For example, for the matrix A above,

$$A' = \begin{bmatrix} 2 & 4 \\ 6 & 7 \\ 3 & 2 \end{bmatrix}$$ (A.20)

A *square matrix* is a matrix where the number of rows is equal to the number of columns. A *diagonal matrix* has non-zero entries only along the main diagonal running from the upper left to the lower right. Diagonal matrices arise most often in applications involving square matrices.

The *identity matrix*, I, is a square, diagonal matrix with ones on the diagonal and zeros everywhere else. It might be thought of as the matrix equivalent of the number "1." Thus $AI = IA = A$.

If A is $n \times n$, its matrix inverse, A^{-1}, is a square matrix that, when pre- or post-multiplying the matrix A, yields the identity matrix (analogous to taking any number, and then multiplying it by its inverse, which yields a product of 1):

$$AA^{-1} = A^{-1}A = I$$ (A.21)

In particular, the 2×2 matrix

$$A = \begin{bmatrix} a & b \\ c & d \end{bmatrix}$$ (A.22)

has inverse

$$A^{-1} = \frac{1}{ad - bc} \begin{bmatrix} d & -b \\ -c & a \end{bmatrix}$$ (A.23)

when $ad - bc$ is not equal to zero. The first term is a scalar, indicating that each of the four elements in the matrix should be premultiplied by it. For example, if

$$A = \begin{bmatrix} 4 & 7 \\ 3 & 9 \end{bmatrix}$$ (A.24)

then the inverse is found by (a) exchanging the elements on the diagonal, and (b) putting a minus sign in front of the elements on the off-diagonal. Then, all of the elements are multiplied by the inverse of the product of the diagonals, minus the product of the off-diagonals. Thus

$$A = \frac{1}{36 - 21} \begin{bmatrix} 9 & -7 \\ -3 & 4 \end{bmatrix} = \begin{bmatrix} 9/15 & -7/15 \\ -3/15 & 4/15 \end{bmatrix}$$ (A.25)

The reader may verify that in this simple example, $AA^{-1} = A^{-1}A = I$. Calculation of matrix inverses for larger square matrices is more complex. Many software packages allow for easy calculation of matrix inverses, including such popular software as Excel.

A.2.3 Matrix Form of Regression

When there are n observations on both a dependent variable y and a set of p independent variables (x), the linear regression model expresses the dependent variable as a linear combination of the x variables:

$$y_1 = \beta_0 + \beta_1 x_{11} + \beta_2 x_{12} + \ldots + \beta_p x_{1p} + \varepsilon_1$$

$$y_2 = \beta_0 + \beta_1 x_{21} + \beta_2 x_{22} + \ldots + \beta_p x_{2p} + \varepsilon_2$$

$$\vdots$$ (A.26)

$$y_n = \beta_0 + \beta_1 x_{n1} + \beta_2 x_{n2} + \ldots + \beta_p x_{np} + \varepsilon_n$$

where x_{ij} refers to observation i for variable j, and ε_i is a disturbance term reflecting the fact that the linear model does not reproduce the y's exactly.

The linear regression equation may also be written in matrix form. This serves at least two purposes – it allows us to write the equation in compact form, and, more importantly, it facilitates the study and understanding of generalizations of linear regression, when for example the underlying assumptions are violated. The matrix form of linear regression is

$$\mathbf{Y} = \mathbf{X}v + \beta$$ (A.27)

where $\mathbf{X}$ is a $n \times (p+1)$ matrix containing ones in the first column, and the values of the p explanatory, independent x-variables in the other columns. $\mathbf{Y}$ is a $n \times 1$ column vector containing the y-values, and β is a $(p+1) \times 1$ column vector containing the value of the intercept as its first element, and the p-values of the coefficients as its other entries. The column vector ε is $n \times 1$, and contains the disturbances, which refer to the unexplained part of the regression model.

Note that the observations for variable i are actually in column $i+1$ of $\mathbf{X}$, since the first column of $\mathbf{X}$ is a column of ones. Each observation corresponds to a row of $\mathbf{X}$, and each column (except for the first) corresponds to one of the explanatory variables.

One advantage of expressing the regression equation in this form is that the estimation of the coefficients (the slopes and the intercept) may then be calculated using a compact and straightforward matrix expression.

To find the regression coefficients, we wish to minimize the sum of squared differences between observed and predicted values (and thus this is referred to as ordinary least squares regression, or OLS). Defining e as the $n \times 1$ vector of differences between observed and predicted values, minimizing the residual sum of squares is equivalent to minimizing

$$\mathbf{e'e} = \left(\mathbf{Y} - \mathbf{X}\hat{\beta}\right)'\left(\mathbf{Y} - \mathbf{X}\hat{\beta}\right) \tag{A.28}$$

through the choice of estimates for the vector of coefficients, β (where the estimate of the unknown β is denoted by $\hat{\beta}$). Note that $\mathbf{e'e}$ is a 1×1 scalar – it is simply equal to the sum of squared residuals. This is a calculus problem that requires (a) finding the derivative of the above expression with respect to $\hat{\beta}$, (b) setting the result from (a) equal to 0, and then (c) solving the resulting expression for $\hat{\beta}$. This leads to a matrix expression for the estimate of $\hat{\beta}$:

$$\hat{\beta} = \left(\mathbf{X'X}\right)^{-1}\mathbf{X'Y} \tag{A.29}$$

where again the "^" is used to denote that this is an estimate of the true value, β. The reader may wish to verify that the order of the matrices on the right-hand side results in a column vector β that has order $(p + 1) \times 1$.

For hypothesis testing, the errors are assumed to have a normal distribution, with variance–covariance matrix equal to $\sigma^2\mathbf{I}$, where $\mathbf{I}$ is $n \times n$. The implications of this are the assumptions of (a) homoscedasticity – that is, all observations have equal variance (since all of the diagonal elements of the variance–covariance matrix are identical), and (b) independence – the zeros on the off-diagonal represent the fact that pairs of observations are independent.

The predicted values are equal to $\hat{\mathbf{Y}} = \mathbf{X}\beta$. The column vector of residuals is the difference between the observed and predicted values, that is, $\mathbf{e} = \mathbf{Y} - \hat{\mathbf{Y}} = \mathbf{Y} - \mathbf{X}\beta$.

In addition to estimating the parameters, we also usually want to test hypotheses. For example, are the coefficients significantly different from zero? To test hypotheses, we need to know something about the distribution of coefficient estimates, when the null hypothesis (most often, the null hypothesis that $\beta = 0$) is true. A key piece of information we need to know is how variable the estimates will be. Certainly different estimates of the coefficients will be derived from different samples, and we need to know how variable these estimates are. The variances of the parameters are given by the diagonal elements of the $(p + 1) \times (p + 1)$ matrix

$$V\left(\hat{\beta}\right) = \hat{\sigma}_{\hat{\beta}}^2 = \hat{\sigma}_e^2\left(\mathbf{X'X}\right)^{-1} \tag{A.30}$$

where the variance of the residuals is

$$\hat{\sigma}_e^2 = \frac{\mathbf{e'e}}{n - p - 1} = \frac{(\mathbf{Y} - \mathbf{X}\beta)'(\mathbf{Y} - \mathbf{X}\beta)}{n - p - 1} \tag{A.31}$$

While the diagonal elements of $V(\hat{\beta})$ give the variances of the individual coefficients, the off-diagonal elements give the covariances between the parameters. It is the variances, and not the covariances, that are by far used the most often in interpretation. For example, a simple t-test for a particular coefficient can be formed by taking the ratio of the estimate to the standard deviation of the residuals (where the latter quantity is often termed the standard error):

$$t = \frac{\hat{\beta} - 0}{\sqrt{\hat{\sigma}_{\hat{\beta}}^2}} = \frac{\hat{\beta}}{\hat{\sigma}_{\hat{\beta}}}$$

(A.32)

A.2.4 Extensions and Generalizations of Ordinary Least Squares Regression

The matrix expression of regression is useful in part because it facilitates the description and understanding of a number of generalizations, including those where one wishes to address unmet assumptions. In this section, we look at five such cases: weighted least squares, generalized least squares, ridge regression, omitted variable bias, and the hat matrix.

A.2.4.1 Weighted Least Squares

When the assumption of homoscedasticity is violated, the variance of the errors along the regression line (or plane) is no longer constant – that is, the variance of the errors can vary with X. In this case of heteroscedasticity, the diagonal elements of the variance–covariance matrix are no longer equal. If we proceed by using ordinary least squares as described above, the estimates will on average be equal to the true value (i.e., they remain unbiased), but they will have inaccurate sampling variances. In this case, we can still estimate the regression parameters through *weighted* least squares, where observations are weighted in inverse proportion to their variance (observations associated with more uncertainty and higher variance are weighted less; observations with lower variance are given more weight). Letting $w_{ii} = 1/\sigma_i^2$, the estimated coefficients are found from

$$\hat{\beta} = (\mathbf{X'WX})^{-1} \mathbf{X'WY}$$

(A.33)

where $\mathbf{W}$ is an $n \times n$ weight matrix with the w_{ii} on the diagonal, and zeros off of the diagonal. The variance–covariance matrix associated with the estimates is now

$$\hat{\sigma}_{\hat{\beta}}^2 = \sigma_e^2 (\mathbf{X'WX})^{-1}$$

(A.34)

Note that some expositions of weighted least squares may set $w_{ii} = \sigma_i^2$; in this case

$$\hat{\beta} = (\mathbf{X'W^{-1}X})^{-1} \mathbf{X'W^{-1}Y}$$

(A.35)

and

$$\hat{\sigma}_{\hat{\beta}}^2 = \hat{\sigma}_e^2 (\mathbf{X'W^{-1}X})^{-1}$$

(A.36)

When the errors are homoscedastic, the diagonal elements of **W** are all equal, and the expression for the variance of the coefficients reduces to $V(\hat{\beta}) = \sigma_{\hat{\beta}}^2 = \sigma_e^2 (\mathbf{X'IX})^{-1} = \sigma_e^2 (\mathbf{X'X})^{-1}$.

A.2.4.2 Generalized Least Squares

In the more general case, we may wish to alter not only the diagonal elements of the variance–covariance matrix, but the off-diagonals as well. This is the case when pairs of observations are correlated (as is usually the case with geographic data, where nearby observations may be positively correlated with one another). In regression analysis with geographic data, the assumption of independent residuals is often violated.

When the form of the dependence is known, denoting the (known) variance–covariance matrix associated with the residuals by **Ω** (the upper case Greek letter "omega"), the regression coefficients can be estimated by

$$\hat{\beta} = (\mathbf{X'\Omega^{-1}X})^{-1} \mathbf{X'\Omega^{-1}Y} \tag{A.37}$$

and the variances and covariances associated with the estimates are given by

$$\hat{\sigma}_{\hat{\beta}}^2 = \hat{\sigma}_e^2 (\mathbf{X'\Omega^{-1}X})^{-1} \tag{A.38}$$

A particular challenge is to specify the matrix **Ω** well. Because of the difficulties associated with estimating **Ω**, individual entries are usually not estimated directly – alternatively, **Ω** is often modeled and parameterized.

A.2.4.3 Ridge Regression

An assumption of multiple linear regression is that there is no substantial or perfect multicollinearity among the independent variables. When multicollinearity exists amongst the explanatory variables, an important consequence is that the variance of the estimates is inflated. The estimates are still unbiased, but they can fluctuate greatly from sample to sample.

There are a number of ways to address multicollinearity. One simple solution is to delete offending variables from the analysis. If two variables are measuring the same thing, and are highly correlated, only one of the two should be used in the regression. The idea behind ridge regression (Hoerl and Kennard, 1970) is to reduce the inflated variance brought about by multicollinearity, at the cost of introducing a little bias in the estimates. Although it may be unsettling to have an estimator that, on average, does not give you the right answer, it can be argued that this is OK – *if* the amount by which it is wrong is offset sufficiently by a reduction in the variability of estimates. Might the estimates be improved by permitting a little bias, if the variance can be reduced substantially?

The *mean square error* is a measure that combines measures of bias and variance to produce an overall measure of the effectiveness of an estimator. In particular, the mean square error is equal to the variance plus the square of the bias. In the case of regression estimates, if the variance can be reduced more than the square of the bias is increased, then ridge regression can improve the mean square error of the estimates (see Figure A.3).

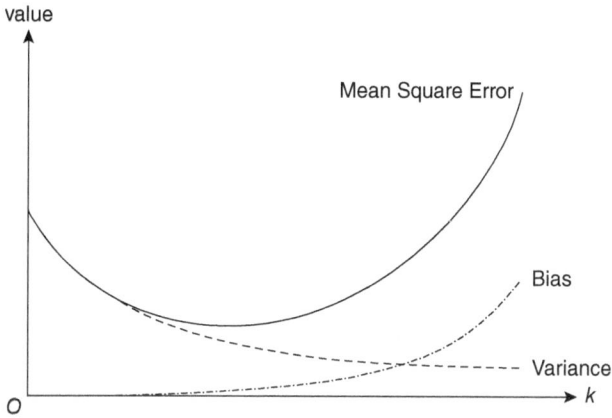

Figure A.3   Bias, variance, and mean square error of regression estimator

The ridge regression estimator is

$$\hat{\beta}_k = (\mathbf{X'X} + k\mathbf{I})^{-1} \mathbf{X'Y} \tag{A.39}$$

where k is a measure of the bias that is permitted (it should be clear that if $k = 0$, the equation reduces to the ordinary least squares (OLS) estimate). Since an effect of multicollinearity is to inflate the variance of the coefficients, by adding k to the diagonal of $\mathbf{X'X}$, the diagonal of the inverse, $(\mathbf{X'X} + k\mathbf{I})^{-1}$ (which is proportional to the variance of the coefficients), will be lower. The idea is that by letting k increase a bit beyond zero, the estimates may be a bit biased, but the variance of the estimates will be lower.

How much should k be increased? How do we know what value of k to use? Hoerl and Kennard (1970) suggest the use of "ridge traces" – for each explanatory, independent variable, β_k is graphed on the vertical axis versus k on the horizontal axis. For variables with a high degree of multicollinearity, these traces will change substantially as k increases from zero and may even change sign; variables with no multicollinearity will exhibit flat traces. The value of k is chosen to be a value where the ridge traces have just begun to flatten out – this is a bit of an art; it is difficult to specify an "optimal" value since the true values of the coefficients are unknown (Figure A.4).

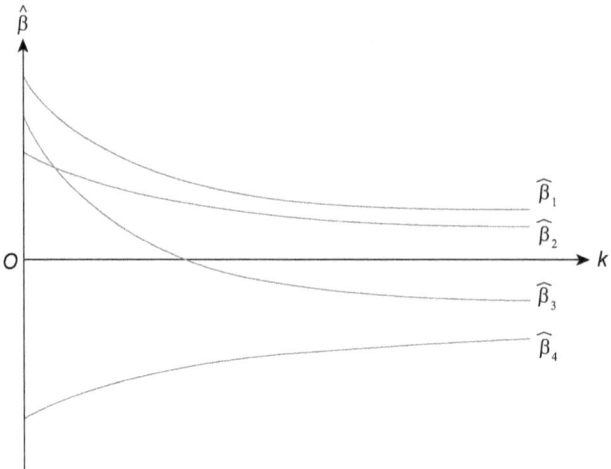

Figure A.4   Ridge traces

A.2.4.4 Omitted Variable Bias

Suppose that a dependent variable, y, depends upon a set of explanatory variables that we can partition into two subsets $\{X_1 , X_2\}$. Further suppose that we have collected information on and included X_1 in our regression, but we have excluded X_2 (because we don't have information on it, or perhaps we don't know that it is important in explaining Y). A consequence is that there will be bias in our estimates of the effects of the X_1 variables on Y. In particular, the expected value of β_1 is equal to

$$E\left[\beta_1\right] = \beta_1 + \mathbf{F}\beta_2 \tag{A.40}$$

where

$$\mathbf{F} = \left(\mathbf{X}_1' \mathbf{X}_1\right)^{-1} \mathbf{X}_1' \mathbf{X}_2 \tag{A.41}$$

The expectation of our estimate is equal to the true value, plus an extra amount, $\mathbf{F}\beta_2$, that is equal to the amount of bias that is introduced. If the true effect of the omitted variables on Y is equal to zero (i.e., $\beta_2 = 0$), there is no omitted variable bias. Note that $\mathbf{F}$ is the set of regression coefficients that result when X_2 is regressed on X_1; that is, $\mathbf{F}$ tells us how X_1 affects X_2. Clearly, if $\mathbf{F} = 0$, then X_1 is not affected by X_2, and there is no bias in the estimates of X_1.

While this is interesting and to some degree informative and insightful, it should be kept in mind that, in practice, we usually do not know what the omitted variables are (if we did, we would have included them – assuming that we had data!). So in practice, we can't actually compute the amount of omitted variable bias, since we either don't know what the variables are, or if we do, we don't have access to the data.

A.2.4.5 Outliers and the Hat Matrix

Outliers can exert substantial effects on regression results. If particular outlying observations are omitted from the analysis, it is possible that the estimated coefficients may be quite different. The hat matrix provides a way to assess the influence that specific observations have on the results. When we substitute $\hat{\beta} = (X'X)^{-1}X'Y$ into the expression for the predicted values of ($\hat{Y} = X\hat{\beta}$), the result is

$$\hat{Y} = X(X'X)^{-1}X'Y = HY \tag{A.42}$$

The hat matrix (alternatively known as the *projection matrix*) $H = X(X'X)^{-1}X'$ is an $n \times n$ matrix. One can see from the equation $\hat{Y} = HY$ that H is a matrix that maps the observed values into the predicted values. Its diagonal elements, h_{ii}, give the influence or "leverage" of observation i on the fit of the regression equation. A general guideline is that if h_{ii} is greater than $2p/n$ (where p is the number of independent variables), observation i should be considered an outlier – the coefficients may change substantially if that observation is omitted. Of course it is not a good practice to simply eliminate outliers to improve fit (e.g., to achieve higher R^2 values), and there should be a good justification for removing observations (i.e., there should be an underlying rationale for why that observation is not representative of the population).

――――――――――――――――――(**Example**)――――――――――――――――――

Four observations on a dependent variable y and an independent variable x are taken:

x	y
1	4
2	6
4	2
5	4

Thus $n = 4$ and $p = 1$.

(a) Find the slope and intercept using the matrix approach to ordinary least squares estimation for linear regression.

First form the matrix **X**; it has four rows (one associated with each observation) and two columns. The first column is a column of ones (the first column of **X** is always a column of ones, and is associated with the intercept; the other columns of **X** are associated with the independent variables), and the second column contains the **X** data:

$$\mathbf{X} = \begin{bmatrix} 1 & 1 \\ 1 & 2 \\ 1 & 4 \\ 1 & 5 \end{bmatrix} \tag{A.43}$$

(Continued)

Next we find that

$$\mathbf{X'X} = \begin{bmatrix} 4 & 12 \\ 12 & 46 \end{bmatrix}; \ (\mathbf{X'X})^{-1} = \begin{bmatrix} 1.15 & -0.3 \\ -0.3 & 0.1 \end{bmatrix}$$

(A.44)

Multiplying this inverse matrix first by $\mathbf{X'}$ and then multiplying the result by $\mathbf{Y}$, we have

$$\hat{\beta} = \begin{bmatrix} 5.2 \\ -0.4 \end{bmatrix}$$

(A.45)

The estimate for the intercept is 5.2, and the estimate for the slope is -0.4. To estimate the variance of the estimates, we first find the estimate of the variance of residuals:

$$\hat{\sigma}_e^{\,2} = \frac{\mathbf{e'e}}{n-p-1} = \frac{\left(\mathbf{Y}-\mathbf{X}\hat{\beta}\right)'\left(\mathbf{Y}-\mathbf{X}\hat{\beta}\right)}{n-p-1} = \frac{6.4}{2} = 3.2$$

(A.46)

The variance of the coefficients is given by the diagonal elements of the variance–covariance matrix $\hat{\sigma}_e^{\,2}(\mathbf{X'X})^{-1}$, and so here we have

$$V\left(\hat{\beta}_0\right) = 3.2(1.15) = 3.68$$

(A.47)

and

$$V\left(\hat{\beta}_1\right) = 3.2(0.1) = 0.32$$

(A.48)

The off-diagonal elements give the covariances:

$$\mathrm{Cov}\left(\hat{\beta}_0, \hat{\beta}_1\right) = 3.2(-0.3) = -0.96$$

(A.49)

The covariances are not used frequently in interpretation, but here we find that there is a negative relationship; the higher the estimate of $\hat{\beta}_0$, the lower the estimate of $\hat{\beta}_1$.

The predicted values are, in matrix form, $\hat{\mathbf{Y}} = \mathbf{X}\hat{\beta}$. We have $\hat{\mathbf{Y}}' = \{4.8, 4.4, 3.6, 3.2\}$ on writing the predictions in a row vector instead of a column vector to save space. The residuals are the differences between observed and predicted values, and thus $\mathbf{e} = \mathbf{Y} - \hat{\mathbf{Y}}$. The vector of residuals is, therefore, $\mathbf{e'} = \{4,6,2,4\} - \{4.8,4.4, 3.6,3.2\} = \{-0.8, 1.6, -1.6, 0.8\}$. Note that the sum of the residuals is equal to zero, as it should be – there are negative and positive residuals and the regression line goes through in a way to have these negative and positive errors "balance out." In matrix terms, we can write the sum of residuals as $\mathbf{1'e}$, where $\mathbf{1}$ is an $n \times 1$ column vector consisting entirely of ones.

The variances may be used to carry out a t-test of the null hypothesis that the coefficients are equal to zero. We are typically most interested in null hypotheses about the slope (and we are often not as concerned with hypotheses about the intercept). We have

$$t = \frac{\hat{\beta}_1 - 0}{\sqrt{V\left[\hat{\beta}_1\right]}} = \frac{\hat{\beta}_1}{\sqrt{V\left[\hat{\beta}_1\right]}} = \frac{-0.4}{\sqrt{0.32}} = 0.566$$

(A.50)

This is to be compared with a critical value of t with $n - 2 = 2$ degrees of freedom. For a Type I error probability of 0.05 and a two-tailed test, the critical values of t are ± 4.303. Since the absolute value of our test statistic is less than 4.303, we do not have enough evidence to reject the null hypothesis.

Exercise

Repeat the illustration above with the following four (x,y) pairs: (3,6), (5,7), (7,12), and (9,7).

Solution

Intermediate results:

$$(\mathbf{X'X})^{-1} = \begin{bmatrix} 2.05 & -0.3 \\ -0.3 & 0.05 \end{bmatrix}$$

$$\hat{\beta}' = [5.6\ 0.4]$$

The intercept is 5.6 and the slope is 0.4. The predicted values are $\hat{\mathbf{Y}} = \mathbf{X}\hat{\beta}$. We find $\hat{\mathbf{Y}}'$ = [6.8, 7.6, 8.4, 9.2]. The residuals are found from $\mathbf{e} = \mathbf{Y} - \hat{\mathbf{Y}}$. The result is $\mathbf{e}' = [-0.8, -0.6, 3.6, -2.2]$. The variance of the residuals is equal to $\hat{\sigma}_e^2 = (\mathbf{e'e})/(n - p - 1) = 18.8/2 = 9.4$ (where $p = 1$ is the number of explanatory variables). This in turn may be used to estimate the variance of the coefficients as $\hat{\sigma}_e^2 (\mathbf{X'X})^{-1} = \begin{bmatrix} 19.27 & -2.82 \\ -2.82 & 0.47 \end{bmatrix}$.

The variance of the intercept and slope appear on the diagonal. We are often interested in the null hypothesis that the slope is equal to zero. To carry out the hypothesis test we find the observed t-statistic as $0.4 / \sqrt{(0.47)}$ and this is less than the critical value of t with $n - 2 = 2$ degrees of freedom. There is not enough information to reject the null hypothesis that the true slope is equal to zero.

A.3 Review and Extension of Some Probability Theory

A discrete random variable, X, has a probability distribution (sometimes called a probability mass function) denoted by $\Pr(X = x) = \Pr(x)$, where x is the value taken by X. A continuous random variable has a probability distribution (also called a probability density function or pdf) denoted by $f(x)$. The likelihood of getting a specific value x is zero, since the distribution is continuous. The likelihood of getting a value within a range $a < x < b$ is equal to the area under the curve, $f(x)$, that lies between a and b. For those familiar with calculus (and even for those not familiar with calculus!), this area is given as the integral of $f(x)$ from a to b:

$$\Pr(a < X < b) = \int_a^b f(x)\, dx \tag{A.51}$$

Again, the integral sign may be thought of as similar to the summation sign; the only difference is that, with a continuous random variable, we have an infinite number of values to sum over. Since the probability of obtaining a value between minus and plus infinity is equal to 1, the total area under the curve f(x) must equal 1:

$$\int_{-\infty}^{+\infty} f(x)\, dx = 1 \tag{A.52}$$

Cumulative distribution functions are denoted with an upper case F, and they tell us the likelihood that the random variable will be less than or equal to a particular value. For a discrete random variable, the probability of obtaining a value less than or equal to a is

$$F(a) = \sum_{x \le a} \Pr(X = x) \tag{A.53}$$

For a continuous random variable, we have

$$F(a) = \Pr(X \le a) = \int_{-\infty}^{a} f(x)\, dx \tag{A.54}$$

A.3.1 Expected Values

The expected value of a random variable, $E[X]$, is also known as the theoretical mean and is denoted by μ. The expected value is given as the weighted average of the possible values the random variable can take on, where the weights are the likelihoods of obtaining those values. For a discrete random variable,

$$E[X] = \mu = \frac{\sum_x x\{\Pr(X = x)\}}{\sum_x \{\Pr(X = x)\}} = \sum_x x\{\Pr(X = x)\} \tag{A.55}$$

For a continuous random variable,

$$E[X] = \mu = \frac{\int_{-\infty}^{+\infty} x f(x)\, dx}{\int_{-\infty}^{+\infty} f(x)\, dx} = \int_{-\infty}^{+\infty} x f(x)\, dx \tag{A.56}$$

In both of these cases, the term in the denominator is equal to 1 (since the sum of all possible outcomes is equal to 1) and the term is written out here to highlight the fact that this is a weighted average – all possible outcomes are weighted by their likelihoods, and then the result is divided by the sum of the weights (which is equal to 1).

Example 1: For the binomial distribution it is possible (although we omit the details) to show that

$$\mu = E[X] = \sum_{x=0}^{x=n} \binom{n}{x} p^x (1-p)^{n-x} = np \tag{A.57}$$

Example 2: If we interview four people, and they each have a 0.1 chance of taking the train to work, the possible outcomes are that we could have 0, 1, 2, 3, or 4 of those people taking the train. Intuitively, 0 would be a fairly likely outcome, and 4 would be an unlikely outcome. The likelihood of each is found by using Equation 2.1: $Pr(X = 0) = 0.9^4 = 0.6560$; $Pr(X = 1) = 4(0.9)^3(0.1) = 0.2916$; $Pr(X = 2) = 6(0.1)^2(0.9)^2 = 0.0486$; $Pr(X = 3) = 4(0.1)^3(0.9) = 0.0036$; and $Pr(X = 4) = 0.1^4 = 0.0001$. Note that these sum to 1; the probability of *some* outcome is equal to 1. In this example, the expected number of people taking the train (i.e., the theoretical mean) is equal to $np = 4(0.1) = 0.4$, and this is the same as $(0)(0.656) + 1(0.2916) + 2(0.0486) + 3(0.0036) + 4(0.0001) = 4(0.1)$.

Example 3: As another example, consider the experiment that consists of rolling a die. What is the expected value of the die? This is equivalent to asking what the expected average is of a large number of rolls:

$$\sum_{x=1}^{6} x\left(\frac{1}{6}\right) = \left(\frac{1}{6}\right)1 + \left(\frac{1}{6}\right)2 + \left(\frac{1}{6}\right)3 + \left(\frac{1}{6}\right)4 + \left(\frac{1}{6}\right)5 + \left(\frac{1}{6}\right)6 = 3.5 \tag{A.58}$$

What is the expected value of a continuous uniform random variable – that is, a random variable that has equally likely outcomes over the range (a, b)? Such a random variable has a probability density function given by

$$f(x) = \frac{1}{(b-a)} \tag{A.59}$$

The expected value is then

$$E[X] = \mu = \int_a^b \frac{x}{(b-a)} dx = \frac{1}{(b-a)}\left(\frac{x^2}{2}\right)\Big|_a^b = \frac{1}{b-a}\frac{b^2-a^2}{2} = \frac{(b-a)(b+a)}{2(b-a)}$$
$$= \frac{a+b}{2} \tag{A.60}$$

The expected value of any function of a random variable, $g(x)$, is a weighted average of the values of $g(x)$, where the weights are again the likelihoods of obtaining the values of $g(x)$. Thus for discrete and continuous variables we have

$$E[g(X)] = \sum_x g(x) Pr(X = x) \tag{A.61}$$

$$E[g(X)] = \int_{-\infty}^{\infty} g(x)f(x)dx \tag{A.62}$$

for discrete and continuous random variables, respectively.

For example, the expected value of the square of a random variable is

$$E[x^2] = \sum_x x^2 Pr(x) \tag{A.63}$$

in the discrete case, and

$$\int_x x^2 f(x) dx \qquad\qquad\qquad\qquad\qquad\text{(A.64)}$$

in the continuous case.

Useful rules for working with expected values are: (i) the expected value of a constant is simply equal to the constant; (ii) the expected value of a constant times a random variable is equal to the constant times the expected value of the random variable; and (iii) the expected value of a sum is equal to the sum of the expected values. These rules are summarized below:

(i) $E[a] = a$ \qquad\qquad\qquad\qquad (A.65)

(ii) $E[bX] = bE[X]$ \qquad\qquad\qquad (A.66)

(iii) $E[a + bX] = a + bE[X]$ \qquad\qquad (A.67)

A.3.2 Variance of a Random Variable

The variance of a random variable, $V[X]$, is the expected value of the squared deviation of an observation from the mean, typically denoted by σ^2:

$$\sigma^2 = V[X] = E[(X - \mu)^2] = E[X^2] - \mu^2 = \sum x^2 \Pr(X = x) - \mu^2 \qquad\qquad\text{(A.68)}$$

Using the rules for expected values above,

$$\begin{aligned} V[X] &= E[X^2 - 2X\mu + \mu^2] = E[X^2] - 2\mu E[X] + \mu^2 \\ &= E[X^2] - \mu^2 \end{aligned} \qquad\qquad\text{(A.69)}$$

For the binomial distribution it may be shown that

$$\sigma^2 = \sum x^2 \Pr(X = x) - \mu^2 = \sum x^2 \sum_{x=0}^{x=n} \binom{n}{x} p^x (1-p)^{n-x} - (np)^2 = np(1-p) \qquad\qquad\text{(A.70)}$$

To take another example, let us return to the experiment involving the roll of a die. The variance of the random variable X in this case is equal to $E[X^2] - 3.5^2$. The expected value of X^2 is found using Equation A.61:

$$E[X^2] = 1^2\left(\frac{1}{6}\right) + 2^2\left(\frac{1}{6}\right) + 3^2\left(\frac{1}{6}\right) + 4^2\left(\frac{1}{6}\right) + 5^2\left(\frac{1}{6}\right) + 6^2\left(\frac{1}{6}\right) = 15.17 \qquad\qquad\text{(A.71)}$$

The variance is therefore equal to $15.17 - 3.5^2 = 2.92$. To illustrate the derivation of the variance using a continuous variable, let us continue with the example of a uniform random variable. We have

$$E\left[X^2\right] = \frac{1}{b-a}\int_a^b x^2 dx = \left(\frac{1}{b-a}\right)\frac{x^3}{3}\bigg|_a^b = \frac{1}{(b-a)}\frac{b^3-a^3}{3} \tag{A.72}$$

Then

$$V[X] = \frac{b^3-a^3}{3(b-a)} - \frac{(a+b)^2}{4} = \frac{(b-a)^2}{12} \tag{A.73}$$

A.4 Parameter Estimation

How do we take a set of data and fit a distribution to that data? If we have some data and would like to fit an exponential distribution, how do we do that? We don't want to choose the parameter λ to be too high, or the fitted curve will decline too steeply. If the estimated value of λ is too low the fitted curve will be too flat.

In this section we describe three ways to estimate the unknown parameters of distributions, given a set of data, $x_1, x_2, x_3, \ldots, x_n$.

Before looking at approaches to parameter estimation, we first examine how we assess estimators. One desirable property for estimators is that they have little or no bias. Bias is the difference between the expected value of the estimator and the true value of the parameter. For a parameter denoted α,

$$Bias[\hat{\alpha}] = E[\hat{\alpha}] - \alpha \tag{A.74}$$

We also want estimators that have a small variance – if we collect other samples, we'd like to get a set of estimates that display little variation and are about the same each time. The variance of an estimator is the expected value of the squared difference between the estimated value and the true value

$$V[\hat{\alpha}] = E\left[(\hat{\alpha} - a)^2\right] \tag{A.75}$$

As we have seen, these two measures are combined by defining the mean square error as being equal to the square of the bias, plus the variance:

$$MSE[\hat{\alpha}] = \left(Bias[\hat{\alpha}]\right)^2 + V[\hat{\alpha}] \tag{A.76}$$

The mean square error is a common way to measure of how good an estimator is.
We now turn to the discussion of various methods of parameter estimation.

A.4.1 Median Estimator

The median estimator is easy to understand and calculate, but (a) it relies on only one data point (the median), and (b) its mean square error is often higher than it is for the other two estimators we will discuss.

The median estimator is found simply by setting the cumulative distribution function for the distribution, $F(x) = \Pr(X < x)$, equal to 0.5 (since half of the observations are less than the median). Then, substituting the observed median for x, we solve for the unknown parameter.

For example, to fit an exponential distribution, we use

$$F(x) = 0.5 = 1 - e^{-\lambda x} \tag{A.77}$$

Substituting the observed median, say x_{med}, for x, we solve for an estimate of λ:

$$e^{-\lambda x_{med}} = 0.5 \tag{A.78}$$

and therefore

$$\ln(0.5) = -\lambda x_{med} \tag{A.79}$$

and

$$\hat{\lambda} = \frac{-\ln(0.5)}{x_{med}} = \frac{0.693}{x_{med}} \tag{A.80}$$

where the " ^ " indicates that we have an estimate of the true, unknown value of λ. As an aside, note that for the positively skewed exponential distribution, the median is less than the mean. In particular, the mean is equal to $1/\lambda$, and the median, as we can deduce from from Equation A.80, is equal to $0.693/\lambda$. For the exponential distribution, the median is always 69.3% of the mean.

A.4.2 Method of Moments

The method of moments is also easy to understand, and often easy to derive. It usually has better properties than the median estimator, in the sense that it often has a lower mean square error.

When estimating a single parameter, the idea behind the method of moments is to set the sample mean equal to the theoretical mean, and then solve for the parameter.

For example, suppose that our random variable has a uniform distribution with known lower limit, $a = 0$, and an unknown upper limit, b. We have four observations, $x_1 = 6$, $x_2 = 8$, $x_3 = 14$, $x_4 = 22$, and wish to estimate the parameter, b.

We proceed by setting the sample mean equal to the theoretical mean:

$$\bar{x} = \frac{a+b}{2} = \frac{b}{2} \tag{A.81}$$

Solving for the parameter,

$$\hat{b} = 2\bar{x} \tag{A.82}$$

For our sample, $\bar{x}$ = 12.5 and $\hat{b}$ = 25.

In some cases, we wish to estimate two parameters. In that case, we need two equations and two unknowns, and the idea is to equate not only the sample mean with the theoretical mean, but also the sample variance with the theoretical variance:

$$\bar{x} = E[X] \tag{A.83}$$

$$s^2 = E[X^2] - \{E[X]\}^2 \tag{A.84}$$

Taking a step back for a "moment," the reader may be wondering – what is a "moment?" The kth moment of a variable X is the expected value of X^k. The kth *central* moment is the expected value of $(X - \mu)^k$.

A.4.3 Maximum Likelihood

Maximum likelihood is another method of parameter estimation. When we have data, and we want to fit it to a distribution, we can use the method of maximum likelihood to do so.

The likelihood of a sample is equal to the probability of obtaining it. The maximum likelihood approach is based upon the idea that parameters should be chosen to make the observed sample as likely as possible. Consider the following example. We interview n = 10 individuals, and ask each whether they take the train to work. Our objective is to estimate the probability of commuting by train (p) in the larger population that resides within the study area. Time and cost considerations prevent us from obtaining the true value of p via enumeration of all individuals. Our sample reveals that x = 2 individuals take the train. Intuitively, we realize that the estimate of the probability of commuting by trains is p = 2/10 = 0.2. How can we formalize this by using the method of maximum likelihood to estimate p? Although we don't need to make something so intuitive and simple more complex (Laplace said that the definition of probability was common sense reduced to calculus!), the benefit of making it more complex here is that we will be able to use the method and apply it to much more complex problems where the solution is not at all intuitive.

The likelihood ($\mathcal{L}$) of the observed sample is, using the binomial distribution, the probability of getting two "successes" in 10 trials:

$$\mathcal{L} = \Pr(X = 2) = \binom{10}{2} p^2 (1-p)^8 = \binom{n}{x} p^x (1-p)^{(n-x)} \tag{A.85}$$

Different values of p will yield different values of $\mathcal{L}$; there are "good" choices of p and there are poor choices of p. For example, if we chose $p = 0.9$, that would make the sample of 2 commuters out of a sample of 10 very unlikely. Since we can choose p to be whatever we like, a good choice would be to seek the value of p (say p^*) that maximizes $\mathcal{L}$ (Figure A.5). In this particular example, it will turn out to be that $\mathcal{L}$ is maximized when $p^* = 0.2$. More generally, the likelihood for binomial experiments like this is proportional to $p^x (1-p)^{n-x}$ (we do not need to be concerned with the leading term on the right-hand side, since it is constant, and not a function of p).

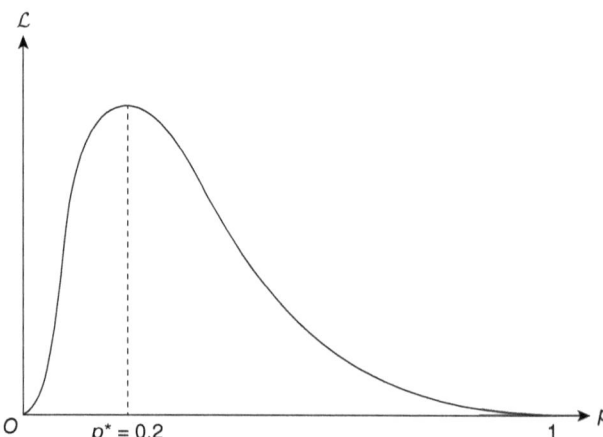

Figure A.5   Likelihood as a function of p

To show that the solution turns out to be $p^* = 0.2$, we need to find the maximum, and for that we solve a calculus problem by finding the point where the slope of the tangent to the likelihood function is equal to zero. More specifically, we take the derivative of the log likelihood with respect to the parameter to be estimated (p). It is almost always easier to work with the (natural) log of the likelihood in seeking the maximum; the natural logarithm of the likelihood is a monotonic transformation, and this implies that the points where both the likelihood function and the log of the likelihood function are maximized are identical. Taking the natural logs of both sides of the likelihood function, we have

$$\ln \mathcal{L} = x \ln p + (n-x) \ln (1-p) \tag{A.86}$$

We are interested in the point where the slope of the graph of $\ln \mathcal{L}$ vs. p is equal to zero, since that corresponds to the point where the log likelihood curve has its maximum. Setting the derivative with respect to p equal to zero we have

$$\frac{\partial \ln L}{\partial p} = \frac{x}{p} - \frac{n-x}{1-p} = 0 \tag{A.87}$$

Solving this for p yields the maximum likelihood estimator, $\hat{p} = x/n$. Although this may be a roundabout way to derive an estimator that is intuitive, the value of the illustration lies in showing the common steps that are followed – write down the likelihood, take the natural log of the likelihood, differentiate this with respect to the unknown parameter (or parameters), set the result equal to zero, and solve for the parameter(s).

More generally, the likelihood of a set of n independent observations, drawn from a common continuous probability distribution $f(x)$, is

$$\mathcal{L} = \prod_{i=1}^{n} f(x_i) \tag{A.88}$$

Another illustration of maximum likelihood estimation is provided by the exponential distribution; this distribution is used in many spatial applications. For instance, spatial interaction is known to decline with distance, and it is of interest to estimate the rate of decline. Data collected on the distance traveled to various destination types (health clinics, for example) often reveal a high frequency of short trips and a low frequency of long trips. If these data are modeled with the exponential distribution,

$$f(x) = \lambda e^{-\lambda x} \tag{A.89}$$

the parameter λ represents the steepness of the decline in interaction with distance (or the "friction" of distance). High values of λ correspond to a steep decay. How do we estimate λ for a set of observed data, when we wish to fit the exponential distribution? We now find the maximum likelihood estimator for λ from a sample of observed distances that are assumed to be independent $\{x_i; i=1, 2, ..., n\}$.

1. The first step is to find the likelihood of the sample:

$$\mathcal{L} = \prod_{i=1}^{n} f(x_i) = \prod_{i=1}^{n} \lambda e^{-\lambda x_i} = \lambda^n \prod_{i=1}^{n} e^{-\lambda x_i} \tag{A.90}$$

(The assumption made here of independent responses allows the individual probabilities to be multiplied together.)

2. Take the natural log of the likelihood:

$$\ln \mathcal{L} = n \ln \lambda - \lambda \sum_{i=1}^{n} x_i \tag{A.91}$$

(This uses the rules that $\ln x^a = a \ln x$, and $\ln(ab) = \ln a + \ln b$.)

3. Set the derivative of the log likelihood with respect to λ equal to zero:

$$\frac{\partial \ln(\mathcal{L})}{\partial \lambda} = \frac{n}{\lambda} - \sum_{i=1}^{n} x_i = 0 \tag{A.92}$$

Solve for λ:

$$\hat{\lambda} = \frac{n}{\sum_{i=1}^{n} x_i} = \frac{1}{\bar{x}} \tag{A.93}$$

The maximum likelihood estimator of λ is simply the reciprocal of the mean. This makes intuitive sense; higher mean distances correspond with lower values of λ, and low values of λ are consistent with small declines of interaction frequencies with distance.

Another example of maximum likelihood is its use in regression. If the residuals in regression are normally distributed about the regression line, the maximum likelihood estimates for the slopes and intercept are also the same as the ordinary least squares estimates.

Maximum likelihood estimates have the property that they are asymptotically unbiased and they are also minimum variance estimates (sometimes abbreviated MVUE, for minimum variance, unbiased estimate). That is, for large sample sizes, the expected value of the estimate will be equal to the true value, and no other estimator will have a lower variance.

A.5 Simulation of Variates from Probability Distributions

Random draws from probability distributions may be simulated by recognizing that the values of the associated cumulative distribution function $F(x)$ are equally likely, and of course range from 0 to 1. Figure A.6 shows the underlying idea. We first choose a value (say u) at random from the uniform distribution ranging from 0 to 1. Setting this equal to $F(x)$, we then solve for the simulated value x^*. Many values chosen in this way will give rise to a histogram or density function that simulates the probability density function, $f(x)$.

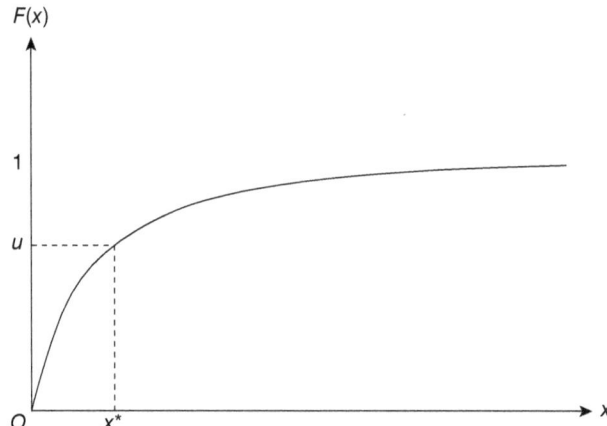

Figure A.6 Simulating variates using the cumulative distribution function

Example. Suppose we wish to simulate a random draw from an exponential distribution with parameter λ. We first set

$$u = U(0,1) = F(x) = 1 - e^{-\lambda x} \tag{A.94}$$

(where $U(0,1)$ is the notation for a uniform variable distributed on the range from 0 to 1), and then solve for x:

$$1 - u = e^{-\lambda x} \tag{A.95}$$

which implies that the simulated value, x^*, is

$$x^* = -\frac{\ln(1-u)}{\lambda} \tag{A.96}$$

A.6 Practice with Distributions

In this section, we will go through a number of examples that highlight and reinforce a number of the ideas we have surveyed in the previous sections.

In particular, we will look at three examples: (1) a contrived example that uses a distribution that is not widely used, but lends itself to illustrating the ideas; (2) the Pareto distribution, which is used often to model incomes, as well other phenomena; and (3) the intervening opportunities model, which is used to model transportation flows.

A.6.1 An Illustration with a Somewhat Contrived Distribution

Suppose that $f(x) = cx^\alpha$ over the range $0 \le x \le 4$, where c is a constant and α is a parameter governing the shape of the probability density function. This is a flexible distribution in the sense that if $\alpha > 1$ the density rises faster than linearly; if $\alpha = 1$ the probability density is linear; and if $\alpha < 1$ the increase is less than linear. All of these examples (where $\alpha > 0$) are examples of negatively skewed distributions, where high values are more likely than low values. Distributions of this form are power-law probability distributions and they are related to the Pareto distribution that we will explore in the next subsection.

We now consider in turn the following tasks:

(a) Find the value of c necessary to make this a valid pdf
(b) Describe how you simulate values from this distribution
(c) Find the theoretical mean (i.e., expected value) and variance

(d) Find the median

(e) Find median, moment, and maximum likelihood estimators for α.

(a) Since the entire area under the curve must be equal to 1, we have

$$\int_0^4 cx^\alpha \, dx = 1 \tag{A.97}$$

This implies that

$$c\frac{x^{\alpha+1}}{\alpha+1}\bigg|_0^4 = \frac{c\left(4^{\alpha+1}\right)}{\alpha+1} - \frac{c\left(0^{\alpha+1}\right)}{\alpha+1} = \frac{c\left(4^{\alpha+1}\right)}{\alpha+1} = 1 \tag{A.98}$$

and therefore

$$c = \frac{\alpha+1}{4^{\alpha+1}} \tag{A.99}$$

(b) To simulate values from the distribution (for a particular value of α), we first need to find $F(x)$. This can be found by integrating the pdf:

$$F(x) = \int_0^x \frac{\alpha+1}{4^{\alpha+1}} x^\alpha \, dx = \frac{\alpha+1}{4^{\alpha+1}} \int_0^x x^\alpha \, dx = \frac{x^{\alpha+1}}{4^{\alpha+1}} = \left(\frac{x}{4}\right)^{\alpha+1} \tag{A.100}$$

Note that $F(0) = 0$ and $F(4) = 1$, as they should. To simulate (i.e., choose a variable from this distribution), first choose a uniform $(0, 1)$ random variate (u). This is equivalent to choosing a percentile for our simulated variate. For example, if we choose $u = 0.73$, we have chosen the 73rd percentile of the variable, and we are looking for the value of x such that $\Pr(X < x) = F(x) = 0.73$. We therefore equate our value of u with $F(x)$:

$$u = \left(\frac{x}{4}\right)^{\alpha+1} \tag{A.101}$$

Now solve this for x; this can be done by raising both sides to the power $1/(\alpha+1)$:

$$x = 4u^{\left(\frac{1}{\alpha+1}\right)} \tag{A.102}$$

Simulating observations from distributions can be useful in studying the distribution because we are creating a controlled experiment where we *know* the true value of the parameter. In comparing various methods for parameter estimation, for example, we can first simulate observations from the known distribution, and then see how well the various methods perform.

(c) The theoretical mean, or expected value, is

$$E[x] = \mu = \int xf(x)\,dx = \frac{\alpha+1}{4^{\alpha+1}} \int_0^4 (x) x^\alpha dx = \frac{\alpha+1}{4^{\alpha+1}} \int_0^4 x^{\alpha+1} dx \tag{A.103}$$

This is evaluated as

$$\frac{\alpha+1}{4^{\alpha+1}} \frac{x^{\alpha+2}}{\alpha+2}\bigg|_0^4 = \left(\frac{\alpha+1}{\alpha+2}\right)\left(\frac{4^{\alpha+2}}{4^{\alpha+1}}\right) = 4\left(\frac{\alpha+1}{\alpha+2}\right) \tag{A.104}$$

The variance is found by first finding the expected value of the square of the variable:

$$E\left[x^2\right] = \frac{\alpha+1}{4^{\alpha+1}}\int_0^4 \left(x^2\right)x^\alpha dx = \frac{\alpha+1}{4^{\alpha+1}}\int_0^4 x^{\alpha+2}dx \qquad\text{(A.105)}$$

and this is evaluated as

$$\left(\frac{\alpha+1}{4^{\alpha+1}}\right)\frac{x^{\alpha+3}}{\alpha+3}\Bigg|_0^4 = \left(\frac{\alpha+1}{4^{\alpha+1}}\right)\frac{4^{\alpha+3}}{\alpha+3} = 16\left(\frac{\alpha+1}{\alpha+3}\right) \qquad\text{(A.106)}$$

The variance is equal to

$$E\left[x^2\right] - \left\{E[x]\right\}^2 = 16\left(\frac{\alpha+1}{\alpha+3}\right) - 16\left(\frac{\alpha+1}{\alpha+2}\right)^2 \qquad\text{(A.107)}$$

(d) The median is found by setting the cdf, $F(x)$, equal to 0.5:

$$F(x) = \left(\frac{x}{4}\right)^{\alpha+1} = 0.5 \qquad\text{(A.108)}$$

which, after solving for x, implies that

$$x_{med} = \frac{4}{2^{\left(\frac{1}{\alpha+1}\right)}} \qquad\text{(A.109)}$$

Now, given a set of data $\{x_1, x_2,..., x_n\}$, how might we estimate the unknown parameter, α? That is, how do we fit a probability distribution to our set of data?

(e) The median, moment, and maximum likelihood estimators of the parameter α are found as follows. For the median estimator, we use the observed median in the cdf and set it equal to 0.5:

$$\left(\frac{x_{med}}{4}\right)^{\alpha+1} = F(x) = 0.5 \qquad\text{(A.110)}$$

Solving this for α by first taking the natural logs of both sides yields

$$(\alpha+1)\ln\left(\frac{x_{med}}{4}\right) = \ln(0.5) \qquad\text{(A.111)}$$

and

$$\hat{\alpha} = \frac{\ln(0.5)}{\ln\left(\frac{x_{med}}{4}\right)} - 1 \qquad\text{(A.112)}$$

For the moment estimator, set the observed mean from the data equal to the expected value (i.e., the theoretical mean):

$$\overline{x} = \frac{4(\alpha+1)}{\alpha+2} \tag{A.113}$$

Solving for α yields the moment estimator:

$$\hat{\alpha} = \frac{2\overline{x}-4}{4-\overline{x}} \tag{A.114}$$

Finally, the maximum likelihood estimator is found by first writing out the likelihood:

$$\mathcal{L} = \prod_{i=1}^{n}\frac{\alpha+1}{4^{\alpha+1}}x_i^{\alpha} = \left(\frac{\alpha+1}{4^{\alpha+1}}\right)^n \prod_{i=1}^{n} x_i^{\alpha} \tag{A.115}$$

The log likelihood is

$$\ln \mathcal{L} = n \ln\left(\frac{\alpha+1}{4^{\alpha+1}}\right) + \alpha\sum \ln x_i = n\{\ln(\alpha+1)-(\alpha+1)\ln(4)\} + \alpha\sum \ln x_i \tag{A.116}$$

Setting the derivative with respect to the parameter α equal to zero,

$$\frac{\partial \ln \mathcal{L}}{\partial \alpha} = n\left\{\frac{1}{\alpha+1}-\ln(4)\right\} + \sum \ln x_i = 0 \tag{A.117}$$

Solving this for α yields

$$\frac{n}{\alpha+1} = n\ln 4 - \sum \ln x_i \tag{A.118}$$

and

$$\hat{\alpha} = \frac{n}{n\ln 4 - \sum \ln x_i} - 1 \tag{A.119}$$

Example

Suppose we have observations 1.3, 2.8, 3.6, 3.8, and 3.9. Note that they are all on the range from 0 to 4, and note that higher values seem more common than lower values (i.e., the distribution is negatively skewed). Estimate α using the three approaches just described.

(a) Median estimator. The observed median is 3.6. The estimate for α is $\{\ln(0.5) / \ln(3.6/4)\} - 1 = 5.579$

(b) Moment estimator. The observed mean is 15.4/5 = 3.08. The moment estimator for α is $\{2(3.08) - 4\}/(4 - 3.08) = 2.16/0.92 = 2.323$

(c) Maximum likelihood estimator. Here we have $\{5/(5 \ln(4) - \sum \ln x_i) - 1 = 2.007$

A.6.2 The Intervening Opportunities Model

The intervening opportunities model was developed by Stouffer (1940) in the context of migration. He developed the model to explain the fact that there were many short-distance moves, and relatively few long-distance moves. Stouffer argued that individuals obeyed the principle of least effort, considering nearby opportunities first. They either accepted them as attractive destinations, or rejected them, moving on to consider the next closest destination. The model assumed that the probability of accepting each destination (denoted here as L) was constant. These ideas have seen most widespread application in the field of transportation, where the ultimate destination of trips coming out of some origin is considered.

If there are D_1 opportunities in the first, closest zone, then the probability that an individual travels beyond the zone is equal to $(1-L)^{D_1}$. Suppose there are D_2 opportunities in the next zone. The probability that the individual goes beyond Zone 2 is equal to the probability of rejecting all of the opportunity in the first two zones, namely $(1-L)^{D_1+D_2}$.

The continuous version of this model (as the zones get smaller in size and greater in number) is as follows. The probability of going beyond Zone 1 is equal to e^{-LD_1}, and the probability of going beyond the first two zones is $e^{-L(D_1+D_2)}$. This is because the quantity $(1-x)^a$ approaches the value e^{-ax} as x gets smaller, and the quantity ax remains constant.

The cumulative distribution function for the continuous model, phrased in terms of distances traveled, represents the probability that a trip has length less than a distance r:

$$F(r) = 1 - e^{-L\rho\pi r^2} \tag{A.120}$$

where opportunity is here assumed to be uniformly distributed with density equal to ρ. Thus the term $\rho\pi r^2$ now takes the place of D, representing the number of opportunities within a distance r of the individual. From this we can determine the probability density function for trips as a function of distance, r; this is accomplished by taking the derivative of $F(r)$ with respect to r:

$$F'(r) = f(r) = 2L\rho\pi r e^{-L\rho\pi r^2} \tag{A.121}$$

We can now write the likelihood associated with a set of observations, $r_1, r_2, ..., r_n$ as

$$\mathcal{L} = \prod_{i=1}^{n} f(r_i) = (2\rho\pi)^n L^n \prod_{i=1}^{n} r_i e^{-L\rho\pi r_i^2} \tag{A.122}$$

The log likelihood is

$$\ln \mathcal{L} = n \ln(2\rho\pi) + n \ln L + \sum_{i=1}^{n} \ln r_i - L\rho\pi \sum_{i=1}^{n} r_i^2 \tag{A.123}$$

Now taking the derivative with respect to L to find the slope, and setting the result equal to 0,

$$\frac{n}{L} - \rho\pi \sum_{i=1}^{n} r_i^2 = 0 \qquad\qquad (A.124)$$

Solving for the parameter L yields the maximum likelihood estimator:

$$L = \frac{n}{\rho\pi \sum_{i=1}^{n} r_i^2} \qquad\qquad (A.125)$$

For the moment estimator of the parameter in the intervening opportunities model, the theoretical mean is set equal to the sample mean:

$$\mu = \int_0^\infty rf(r)dr = \int_0^\infty 2\rho\pi L r^2 e^{-L\rho\pi r^2}\, dr = \frac{1}{2\sqrt{\rho L}} = \bar{r} \qquad\qquad (A.126)$$

Solving for the parameter L yields

$$L = \frac{1}{4\rho\bar{r}^2} \qquad\qquad (A.127)$$

Yet another way to estimate the unknown parameter is to use the median estimator. Similar to the method of moments, we set the sample median equal to the theoretical median; the latter is achieved by setting the cumulative distribution function equal to one-half (since for the median, the chance of observing a lower value is equal to one-half):

$$F(x_{med}) = 0.5 \qquad\qquad (A.128)$$

where x_{med} refers to the observed median.

For the intervening opportunities model, we have a set of distances, $r_1, r_2, ..., r_n$. Denote the observed median among this set as r_{med}. Now set the cumulative distribution function, $F(r_{med})$ equal to 0.5:

$$1 - e^{-L\rho\pi r_{med}^2} = 0.5 \qquad\qquad (A.129)$$

and solving for the unknown parameter,

$$L = \frac{-\ln(0.5)}{\rho\pi r_{med}^2} \qquad\qquad (A.130)$$

Often the median estimators are easiest to define, but unfortunately they sometimes do not have the best performance – for example, their variances may be high.

Comparative Performance

Which estimator should we use? For example, is one estimator better than the others when we have small samples? We have made some general statements – maximum likelihood estimators are asymptotically unbiased; median estimators are often easy to derive, but they tend to underperform.

We can investigate this question in more detail through simulation experiments conducted in the following way:

First choose a sample size, n. This could be repeated for a number of values of n; typically we will want to look at small values of n since we already know that the maximum likelihood estimator is the minimum variance unbiased estimator for large n. Next, choose a value for the unknown parameter (or values for the unknown parameters, if there is more than one). This is where we have some control and this is what provides the rationale for simulation – with an actual dataset, we only have a sample, and we don't know the true values of the parameters. Then we simulate data from the distribution and estimate the parameters. This is all repeated a large number of times – say, 10,000. Now we have a distribution of parameter values, and we can compare that distribution to the known value of the parameter. In particular we are able to compute the bias in parameter estimates, the variance of the estimates, and the associated mean square error.

How can we simulate observations for this intervening opportunities model? We begin by assuming that we know the value of the parameter L. Next, we draw samples of distances traveled that have the probability density function described by Equation A.121. A simulated observation (r) from a known probability distribution can be taken by setting a uniform random variable between 0 and 1 (say, u) equal to the desired cumulative density function, $F(r)$,

$$u = F(x) \tag{A.131}$$

and then solving for x. For the intervening opportunities model,

$$u = 1 - e^{-L\rho\pi r^2} \tag{A.132}$$

and solving this for r:

$$r = \sqrt{\frac{-\ln(1-u)}{L\rho\pi}} \tag{A.133}$$

As noted in Section A.5, this procedure of simulating random variables is a general one. All we need to do is choose a random number from a uniform distribution in the range (0,1), set it equal to the cumulative density function, and then solve for the simulated value.

Rogerson (1986) has evaluated alternative estimators for the intervening opportunities model as described above, and we now summarize this approach and findings. The important aspect here is the approach, and not the findings – our purpose here is to lay out the ideas, which are very general – they apply not only to this example but to evaluation of estimators in general.

Suppose we choose $L = 0.01$, and then take $n = 11$ random observations from the distribution in Equation A.120. Without loss of generality, we also assume for simplicity that $\rho = 1$. We can use maximum likelihood, moment, and median approaches to finding estimates for L, based on the 11 observations. If we repeat this many times, we would expect and/or hope that the mean estimated value of L would turn out to be near its true value of 0.01. Furthermore, we would hope that repeated experiments would not only have an average estimate that is near the true value, but also that the variability of the estimates would not be too large – estimates that fluctuate wildly would not be desirable.

Rogerson (1986) finds the results that are displayed in Table A.3. The table is based on 10,000 repetitions (e.g., $n = 11$ observations are used to estimate L, and then this is repeated 10,000 times, so that we have a total of 10,000 estimates of L for each estimator). The table shows that, for small samples (when $n = 11$), all three of the estimators have means that are close to the true value of $L = 0.01$. All are slightly higher than 0.01, and the median estimator has the largest bias (i.e., it is furthest from the true value). The variance of the maximum likelihood and moment estimators are lower than the variance associated with the median estimator.

When $n = 11$, the moment estimator has a mean square error that is slightly lower than the maximum likelihood estimator; the median estimator is clearly inferior. For the case of the larger sample size ($n = 21$), the mean square error is now lowest for the maximum likelihood estimator; again the median estimator has the largest mean square error.

Table A.3 Mean, variance, and mean square error of three estimators

	Maximum likelihood	Moment	Median
$n = 11$			
Mean of L	0.01096	0.01085	0.01146
Variance of L	1.432×10^{-5}	1.367×10^{-5}	3.527×10^{-5}
Mean square error	1.523×10^{-5}	1.439×10^{-5}	3.740×10^{-5}
$n = 21$			
Mean of L	0.01036	0.01049	0.01066
Variance of L	5.53×10^{-6}	6.449×10^{-5}	1.224×10^{-5}
Mean square error	5.66×10^{-6}	6.685×10^{-6}	12.66×10^{-6}

Source: Rogerson (1986). Reprinted with permission of John Wiley and Sons.

A.6.3 Pareto Distribution

The Pareto distribution is a positively skewed distribution. It is commonly used to model income inequality, and it is also used to model such diverse phenomena as city size distributions and the distribution of maximum daily precipitation values observed each year.

It is a relatively easy distribution to work with, and hence lends itself as an illustration of the ideas being discussed here.

The cumulative distribution function is

$$F(x) = 1 - x^{-q}; q > 0; x > 1 \qquad \text{(A.134)}$$

and the probability density function is

$$f(x) = qx^{-(q+1)} \qquad \text{(A.135)}$$

There is one parameter, q, that governs the shape of the density function.

The theoretical mean associated with the distribution is found as follows (for the case where $q > 1$):

$$E[X] = \mu = \int_1^\infty xf(x)dx = \int_1^\infty qx^{-q}dx = \frac{q}{q-1} \qquad \text{(A.136)}$$

We now proceed to find the median, moment, and maximum likelihood estimators for the parameter q. Suppose that we have a set of observations, $x_1, x_2, ..., x_n$. By definition, the median value is the one at which $\Pr(X<x) = F(x) = 0.5$:

$$F(x_{med}) = 0.5 = 1 - x_{med}^{-q} \qquad \text{(A.137)}$$

where x_{med} denotes the observed median. Solving for q yields the median estimator:

$$\hat{q} = \frac{-\ln 0.5}{\ln(x_{med})} = \frac{0.693}{\ln(x_{med})} \qquad \text{(A.138)}$$

For the moment estimator we set the sample mean equal to the theoretical mean (found above):

$$\bar{x} = \frac{q}{q-1} \qquad \text{(A.139)}$$

and we find the moment estimator by solving this for q:

$$\hat{q} = \frac{\bar{x}}{\bar{x}-1} \qquad \text{(A.140)}$$

Finally, we now derive the maximum likelihood estimator. We begin by writing the likelihood as the product of the probability density function values associated with each individual observation. Taking the product assumes that the observations are independent:

$$\mathcal{L} = \prod_{i=1}^{n} f(x_i) = \prod_{i=1}^{n} qx_i^{-(q+1)} = q^n \prod_{i=1}^{n} x_i^{-(q+1)} \qquad \text{(A.141)}$$

We wish to find the value of the parameter q that maximizes this likelihood. Since it almost always proves easier to maximize the log of the likelihood, we find the log likelihood:

$$\ln \mathcal{L} = n \ln q - (q+1) \sum_{i=1}^{n} \ln x_i \tag{A.142}$$

The maximum is found by finding the point where the slope, or derivative, of this curve is equal to zero:

$$\frac{\partial \ln \mathcal{L}}{\partial q} = \frac{n}{q} - \sum_{i=1}^{n} \ln x_i = 0 \tag{A.143}$$

Solving this for q yields

$$\hat{q} = \frac{n}{\sum_{i=1}^{n} \ln x_i} = \left(\overline{\ln x}\right)^{-1} \tag{A.144}$$

Thus the maximum likelihood estimator is equal to the reciprocal of the average of the logged observations (using natural logs).

Simulating observations that have a Pareto distribution (e.g., to carry out a comparison of the three estimators, as was done in the previous subsection which focused on the intervening opportunities model) is achieved by setting a uniform random variate with range $(0,1)$ equal to the cumulative distribution function:

$$u = F(x) = 1 - x^{-q} \tag{A.145}$$

We wish to solve for x. This is carried out in a few algebraic steps:

$$1 - u = x^{-q} \tag{A.146}$$

and

$$(1-u)^{-1} = x^q \tag{A.147}$$

which implies that

$$x = \left(\frac{1}{1-u}\right)^{\frac{1}{q}} \tag{A.148}$$

A simulation experiment was carried out, where n observations were chosen from a Pareto distribution that had parameter $q = 1.5$. Each of the three methods of estimation was used to estimate the value of q. This was repeated 10,000 times, and the mean of the 10,000 repetitions was noted for each estimator. This entire process was repeated three times. Table A.4 shows that the estimator with the least bias is the maximum likelihood estimator; the median estimator has the most bias. For each estimator, there is variability in the estimator. If the number of simulations were increased, for example, from 10,000 to 100,000, the variability across the three repetitions for each of the estimators would be less.

Table A.4 Parameter estimates for Pareto distribution with $q = 1.5$

Median: {1.434, 1.511, 1.435}
Moment: {1.536, 1.521, 1.476}
Maximum likelihood: {1.490, 1.495, 1.508}

Exercises

1. Solve the following integrals:

    (a) $\int_{1}^{2} e^{x} dx$

    (b) $\int_{4}^{5} 0.3 x^{2} dx$

    (c) $\int_{1}^{3} 3\left(x^{2} - x^{1.5}\right) dx$

2. Find derivatives for the following:
    (a) $2x$
    (b) $4x^{5} - 3x^{2}$
    (c) $4.2e^{x}$
    (d) $7e^{(2x)}$

3. Let $\mathbf{A} = \begin{bmatrix} 4 & 5 & 6 \\ 10 & 6 & 7 \end{bmatrix}$; $\mathbf{B} = \begin{bmatrix} 1 & 2 \\ 2 & 5 \\ 3 & 6 \\ 5 & 7 \end{bmatrix}$

    (a) Find the product **AB**.
    (b) Find $(\mathbf{AB})^{-1}$.

4. A discrete random variable takes on the values 0, 1, 2, or 3 with the probabilities 0.3, 0.4, 0.1, and 0.2, respectively. Find the expected value and the theoretical variance.

5. A continuous random variable has the probability density function $f(x) = cx^{0.3}$, with a range from 0 to 3.

    (a) Find the value of c that is required to make this a valid probability density function.
    (b) Find the expected value of the random variable.
    (c) Find the variance of the random variable.
    (d) Find the cumulative distribution function.
    (e) Use the cumulative distribution function to describe how you would simulate observations from this distribution, and then choose five random variates from it.

6. The cumulative distribution function for the Weibull distribution is

$$F(x) = 1 - e^{-(x/\lambda)^k}$$

where the two parameters are λ and k.

a. Suppose λ is known. What is the median estimator for k?

7. Simulate ten 1 observations from a Pareto distribution with parameter $q = 1.3$, and then use moment, median, and maximum likelihood estimators to estimate the parameter from the sample.

Online resources

@

The website's "Further Resources" section contains links to other material which supplements and expands upon information presented in Appendix A. A set of conceptual questions and targeted exercises are also available to help the student review key concepts.

APPENDIX B

Equations for Azimuthal Equidistant Projection

Equations for azimuthal equidistant projection (Snyder, 1987)

(a) To convert a latitude–longitude pair (φ, γ) to x- and y-coordinates when projection is centered on (φ_1, γ_1):

$$x = Rk' \cos\varphi \sin(\gamma - \gamma_1)$$

$$y = Rk' \left[\cos\varphi_1 \sin\varphi - \sin\varphi_1 \cos\varphi \cos(\gamma - \gamma_1) \right]$$

where $k' = c/\sin c$, and

$$\cos c = \sin\varphi_1 \sin\varphi + \cos\varphi_1 \cos\varphi \cos(\gamma - \gamma_1)$$

(b) To convert an x–y coordinate pair to latitude–longitude:

$$\varphi = \arcsin\left[\cos c \sin\varphi_1 + y \sin c \cos\varphi_1 / c \right]$$

$$\gamma = \gamma_1 + \arctan\left[x \sin c / (c \cos\varphi_1 \cos c - y \sin\varphi_1 \sin c) \right]$$

References

Abbott, E.J. and Firestone, F.A. (1933) 'Specifying surface quality: A method based on accurate measurement and comparison', *Mechanical Engineering*, 55: 569–72.

Abofadel, E. and Austin, D. (2006) 'A new method for computing the mean center of population of the United States', *Professional Geographer*, 58: 65–9.

Anselin, L. (1988) *Spatial Econometrics: Methods and Models*. Dordrecht: Kluwer Academic.

Anselin, L. (1995) 'Local indicators of spatial association – LISA', *Geographical Analysis*, 27: 93–115.

Anselin, L. (2009) 'Spatial regression', in A.S. Fotheringham and P.A. Rogerson (eds), *The SAGE Handbook of Spatial Analysis*. London: Sage. pp. 255–76.

Arachcige, C.N.P.G., Prendergast, L.A., and Staudte, R.G. (2019) 'Robust analogues to the coefficient of variation', https://arxiv.org/pdf/1907.01110.pdf (accessed September 12, 2020).

Bachi, R. (1963) 'Standard distance measures and related methods for spatial analysis', *Papers of the Regional Science Association*, 10: 83–132.

Bailey, T. and Gatrell, A. (1995) *Interactive Spatial Data Analysis*. Harlow: Longman.

Barmore, F.E. (1993) 'Where are we? Comments on the concept of "center of population"', *The Wisconsin Geographer*, 9: 8–21.

Bashein, G. and Detmer, P.R. (1994) 'Centroid of a polygon', in P.S. Heckbert (ed.), *Graphics Gems IV*. San Diego, CA: Morgan Kaufman. pp. 3–6.

Benjamini, Y. (2010) 'Discovering the false discovery rate', *Journal of the Royal Statistical Society Series B*, 72(4): 405–16.

Benjamini, Y. and Hochberg, Y. (1995) 'Controlling the false discovery rate: A practical and powerful approach to multiple testing', *Journal of the Royal Statistical Society Series B*, 57(1): 289–300.

Besag, J. and Newell, J. (1991) 'The detection of clusters in rare diseases', *Journal of the Royal Statistical Society, Series A*, 154: 143–55.

Bivand, R.S., Pebesma, E., and Gomez-Rubio, V. (2013) *Applied Spatial Data Analysis with R* (2nd edition). New York, NY: Springer.

Blateyron, F. (2013) 'The areal field parameters', in R. Leach (ed.), *Characterisation of Areal Surface Texture*. Berlin: Springer. pp. 15–43.

Boulesteix, A.-L. (2006) 'Maximally selected chi-square statistics for ordinal variables', *Biometrical Journal*, 48: 451–62.

Brunsdon, C. and Comber, L. (2015) *An Introduction to R for Spatial Analysis and Mapping*. London: Sage.

Burt, J.E., Barber, G.M., and Rigby, D.L. (2020) *Elementary Statistics for Geographers* (3rd edition). New York, NY: Guilford Press.

Ceriani, L. and Verme, P. (2012) 'The origins of the Gini index: Extracts from Variabilitia e Mutabilita (1912) by Corrado Gini', *Journal of Economic Inequality*, 10: 421–43.

Chun, Y. and Griffith, D.A. (2013) *Spatial Statistics and Geostatistics: Theory and Applications for Geographic Information Science and Technology*. London: Sage.

Clark, P.J. and Evans, F.C. (1954) 'Distance to nearest neighbor as a measure of spatial relationships in populations', *Ecology*, 35: 445–53.

Conover, W.J. (1972) 'A Kolmogorov goodness-of-fit test for discontinuous distributions', *Journal of the American Statistical Association*, 67: 591–96.

Cressie, N.A.C. (2015) *Statistics for Spatial Data* (revised edition). New York, NY: Wiley.

Cressie, N.A.C. and Wikle, C.K. (2011) *Statistics for Spatio-Temporal Data*. Hoboken, NJ: Wiley.

Curtis, J. and McIntosh, R. (1950) 'The interrelationships of certain analytic and synthetic phytosociological characters', *Ecology*, 31: 434–55.

Cuzick, J. and Edwards, R. (1990) 'Spatial clustering for inhomogeneous populations', *Journal of the Royal Statistical Society Series B*, 52: 73–104.

Cysarz, D., Edelhäuser, F., and Van Leeuwen, P. (2015) 'Strategies of symbolization in cardiovascular time series to test individual gestational development in the fetus', *Philosophical Transactions of the Royal Society A: Mathematical, Physical and Engineering Sciences*, 373: 20140087. doi.org/10.1098/rsta.2014.0087

Dale, M.R.T. and Fortin. M.-J. (2014) *Spatial Analysis: A Guide for Ecologists* (2nd edition). Cambridge: Cambridge University Press.

Douglas, E. (1930) 'Boundaries, Area, Geographical Centers, and Altitudes of the United States and the Several States', *Bulletin 817, Geological Survey*. Washington, DC: US Government Printing Office.

Duncan, O.D., Cuzzort, R.P., and Duncan, B. (1961) *Statistical Geography: Problems in Analyzing Areal Data*. Glencoe, IL: The Free Press.

Fischer, M.M. and Getis, A. (2009) *Handbook of Applied Spatial Analysis: Software Tools, Methods, and Applications*. Berlin: Springer.

Flood, S., King, M., Rodgers, R., Ruggles, S., and Warren, J.R. (2020) Integrated Public Use Microdata Series, Current Population Survey: Version 8.0 [dataset]. Minneapolis, MN: IPUMS, 2020. https://doi.org/10.18128/D030.V8.0

Fotheringham, A.S. and Zhan, F.B. (1996) 'A comparison of three exploratory methods for cluster detection in point patterns', *Geographical Analysis*, 28: 200–18.

Fuchs, C. and Kennet, R. (1980) 'A test for detecting outlying cells in the multinomial distribution and two-way contingency tables', *Journal of the American Statistical Association*, 75: 395–8.

Geary, R.C. (1954) 'The contiguity ratio and statistical mapping', *The Incorporated Statistician*, 5(3): 115–45.

Gelfand, A.E., Diggle, P.J., Fuentes, M., and Guttorp, P. (2010) *Handbook of Spatial Statistics*. Boca Raton, FL: Chapman & Hall/CRC Press.

Gelman, A. and Nolan, D. (2017) *Teaching Statistics: A Bag of Tricks*. New York, NY: Oxford University Press.

Getis, A. and Ord, K. (1992) 'The analysis of spatial association by use of distance statistics', *Geographical Analysis*, 24: 189–206.

Gini, C. (1912) 'Variabilità e Mutabilità: Contributo allo studio delle distribuzioni e delle relazioni statistiche', *Studi economico-giuridici Anno III, Parte II*. Bologna: Cuppini.

Gini, C. (1914) 'Sulla misura della concentrazione e della variabilità dei caratter', *Atti del Reale Istituto Veneto di Scienze, Lettere ed Arti*, 73(2): 1203–48.

Gleason, H.A. (1920) 'Some applications of the quadrat method', *Bulletin of the Torrey Botanical Club* 47(1): 21–33.

Gong, J. (2002) 'Clarifying the standard deviational ellipse', *Geographical Analysis*, 34(2): 155–67.

Gould, S.J. (1992) *Bully for Brontosaurus: Reflections in Natural History*. New York, NY: W.W. Norton.

Griffith, D. (2005) 'Effective geographic sample size in the presence of spatial autocorrelation', *Annals of the Association of American Geographers*, 95(4): 740–60.

Griffith, D. (2019) 'Negative spatial autocorrelation: One of the most neglected concepts in spatial statistics', *Stats*, 2(3): 388–415.

Grohmann, C.H., Smith, M.J., and Riccomini, C. (2011) 'Multi-scale analysis of topographic surface roughness in the Midland Valley, Scotland', *IEEE Transactions on Geoscience and Remote Sensing*, 49: 1200–13.

Haining, R. (1988) 'Estimating spatial means with an application to remotely sensed data', *Communications in Statistics: Theory and Methods*, 17(2): 573–97.

Haining, R. (2003) *Spatial Data Analysis: Theory and Practice*. Cambridge: Cambridge University Press.

Han, D. and Rogerson, P. (2003) 'Application of a GIS-based statistical method to assess spatio-temporal changes in breast cancer clustering in the northeastern United States', in O. Khan and R. Skinner (eds), *Geographic Information Systems and Health Applications*. Hershey, PA: Idea Group. pp. 114–38.

Han, D., Carrow, S.S., Rogerson, P.A., and Munschauer, F.E. (2005) 'Geographical variation in cerebrovascular disease in New York State: The correlation with income', *International Journal of Health Geographics*, 4: 25.

Harlin, J. (1978) 'Statistical moments of the hypsometric curve and its density function', *Journal of the International Association for Mathematical Geology*, 10(1): 59–72.

Henderson, R. (1986) 'Change-point problem with autocorrelated observations, with an application to material accountancy', *Technometrics*, 28: 381–9.

Hoerl, A.E. and Kennard, R.W. (1970) 'Ridge regression: Biased estimation for nonorthogonal problems', *Technometrics*, 12(1): 55–67.

Hoover, E. (1941) 'Interstate redistribution of population: 1850–1940', *Journal of Economic History*, 1: 199–205.

James, B., James, K.L., and Siegmund, D. (1987) 'Tests for a changepoint', *Biometrika*, 74: 71–83.

Katz, I.N. and Cooper, L. (1980) 'Optimal location on a sphere', *Computers and Mathematics with Applications*, 6: 175–96.

Koehler, K.J. and Larntz, K. (1980) 'An empirical investigation of goodness-of-fit statistics for sparse multinomials', *Journal of the American Statistical Association*, 75: 336–44.

Koziol, J.A. (1991) 'On maximally selected chi-square statistics', *Biometrics*, 47: 1557–61.

Kulldorff, M. (1997) 'A spatial scan statistics', *Communications in Statistics: Theory and Methods*, 26(6): 1481–96.

Kulldorff, M. and Nagarwalla, N. (1995) 'Spatial disease clusters: Detection and inference', *Statistics in Medicine*, 14(8): 799–810.

Lefever, D.W. (1926) 'Measuring geographic concentration by means of the standard deviational ellipse', *The American Journal of Sociology*, 32(1): 88–94.

Leighton, R. (1991) *Tuva or Bust! Richard Feynman's Last Journey*. New York, NY: W.W. Norton.

Leonov, N.I. (1927) Танну-Тува страна голубой реки *[Tannu Tuva: Land of the Blue River]*. Moscow: Изд-во Об-ва политкаторжан [Izd-vo Ob-va politkatorzhan].

LeSage, J. and Pace, R.K. (2009) *Introduction to Spatial Econometrics*. Boca Raton, FL: Chapman & Hall/CRC Press.

Li, H., Calder, C., and Cressie, N. (2007) 'Beyond Moran's *I*: Testing for spatial dependence based on the spatial autoregressive model', *Geographical Analysis*, 39: 357–75.

Li, H., Calder, C., and Cressie, N. (2012) 'One-step estimation of spatial dependence parameters: Properties and extensions of the APLE statistic', *Journal of Multivariate Analysis*, 105: 68–84.

Litwhiler, D.W. (1976) 'Large region location problems'. PhD dissertation. University of Oklahoma, Norman, OK.

Litwhiler, D.W. and Aly, A.A. (1979) 'Large region location problems', *Computers and Operations Research*, 6: 1–12.

Long, L. and Nucci, A. (1997) 'The Hoover index of concentration: A correction and an update', *Professional Geographer*, 49(4): 431–40.

Lorenz, M.O. (1905) 'Methods of measuring the concentration of wealth', *Publications of the American Statistical Association*, 9(70): 209–19.

Mariappan, P. (2013) *Biostatistics*. Noida, Uttar Pradesh: Pearson India.

McCollum, J.M. (2011) 'Tug of war at the border', *Cartographica*, 46(2): 127–33. doi.org/10.3138/carto.46.2.127

McDonald, J.H. (2014) 'G-test of goodness-of-fit', in J.H. McDonald (ed.), *Handbook of Biological Statistics* (3rd edition). Baltimore, MA: Sparky House. pp. 53–8.

McGrew, Jr., J.C., Lembo, Jr., A.J., and Monroe, C.B. (2014) *An Introduction to Statistical Problem Solving in Geography* (3rd edition). Long Grove, IL: Waveland Press.

Melton, M.A. (1965) 'The geomorphic and paleoclimatic significance of alluvial deposits in Southern Arizona', *Journal of Geology*, 73: 1–38.

Mercer, W.B. and Hall, A.D. (1911) 'The experimental error of field trials', *Journal of Agricultural Science*, 4: 107–32.

Miller, R. and Siegmund, D. (1982) 'Maximally selected chi-square statistics', *Biometrics*, 48: 1011–16.

Moran, P.A.P. (1950) 'Notes on continuous stochastic phenomena', *Biometrika*, 37: 17–23.

Naughton B. (2007) *The Chinese Economy: Transitions and Growth*. Cambridge, MA: MIT Press. pp. 18–19.

Olaya, V. (2009) 'Basic land-surface parameters', in T. Hengl and H.I. Reuter (eds), *Geomorphometry: Concepts, Software, Applications*. Volume 33 of *Developments in Soil Science*. Amsterdam: Elsevier. pp. 141–69.

Openshaw, S., Charlton, M., Wymer, C., and Craft, A. (1987) 'A Mark I geographical analysis machine for the automated analysis of point data sets', *International Journal of Geographical Information Systems*, 1: 335–58.

Ord, J. and Getis, A. (1995) 'Local spatial autocorrelation statistics: Distributional issues and an application', *Geographical Analysis*, 27: 286–306.

Ord, K. (1975) 'Estimation methods for models of spatial interaction', *Journal of the American Statistical Association*, 70: 120–6.

Pike, R. and Wilson, S. (1971) 'Elevation-relief ratio, hypsometric integral, and geomorphic area-altitude analysis', *Geological Society of America Bulletin*, 82: 1079–84.

Pinker, S. (2011) *The Better Angels of Our Nature: Why Violence Has Declined*. New York, NY: Viking Books.

Plane, D. and Rogerson, P. (2015) 'On disaggregating and tracking center points of population', *Annals of the Association of American Geographers*, 105(5): 968–86.

Promptov, Y. (1950) В центре азиатского континента [In the center of the Asian continent]. Moscow: Госкультпросветиздата [Goskultprosvetizdata].

Rogerson, P. (1986) 'Parameter estimation in the intervening opportunities model', *Geographical Analysis*, 18(4): 357–60.

Rogerson, P. (1999) 'The detection of clusters using a spatial version of the chi-square goodness-of-fit statistic', *Geographical Analysis*, 31: 130–47.

Rogerson, P. (2001) 'A statistical method for the detection of geographic clustering', *Geographical Analysis*, 33: 215–27.

Rogerson, P. (2006) 'Statistical methods for the detection of clustering in case-control data', *Statistics in Medicine*, 25(5): 811–23.

Rogerson, P. (2011) 'Optimal geographic scales for local statistics', *Statistical Methods in Medical Research*, 20: 119–129.

Rogerson, P. (2012) 'Weight estimation and significance testing for three focused statistics', *Statistical Methods in Medical Research*, 21(5): 433–44.

Rogerson, P. (2013) 'The Gini coefficient of inequality: A new interpretation', *Letters in Spatial and Resource Sciences*, 6(3): 109–120. doi.org/10.1007/s12076-013-0091-x

Rogerson, P. (2015a) 'A new method for finding geographic centers, with application to U.S. states', *The Professional Geographer*, 67: 686–94.

Rogerson, P. (2015b) 'The maximum Getis-Ord statistic adjusted for spatially autocorrelated data', *Geographical Analysis*, 47: 20–33.

Rogerson, P. (2019) 'I dream of Gini: Measures of population concentration and their application to US population concentration', in R. Franklin (ed.), *Population, Place, and Spatial Interaction: Essays in Honor of David Plane*. Singapore: Springer. pp. 1–17.

Rogerson, P. (2020) *Statistical Methods for Geography: A Student's Guide* (5th edition). London: Sage.

Rogerson, P., Sinha, G., and Han, D. (2006) 'Recent changes in the spatial pattern of prostate cancer in the United States', *American Journal of Preventive Medicine*, 30: S50–9.

Rogerson, P. and Wang, L. (2013) 'Simple scan tests for spatial clustering on a square lattice', *Geographical Analysis*, 46: 202–11.

Rogerson, P. and Yamada, I. (2009) *Statistical Detection and Surveillance of Geographic Clusters*. Boca Raton, FL: CRC.

Rushton, G. and Lolonis, P. (1996) 'Exploratory spatial analysis of birth defects in an urban population', *Statistics in Medicine*, 23: 2195–214.

Salinas, M., López-Garrigós, M., Flores, E., Uris, J., and Leiva-Salinas, C. (2015) 'Larger differences in utilization of rarely requested tests in primary care in Spain', *Biochemia Medica*, 25(3): 410–15. doi. org/10.11613/BM.2015.041

Sasvári, Z. and Chen, H. (1999) 'Tight bounds for the normal distribution', *American Mathematical Monthly*, 106: 76.

Schabenberger, O. and Gotway, C.A. (2005) *Statistical Methods for Spatial Data Analysis*. Boca Raton, FL: Chapman & Hall/CRC.

Siegmund, D.O. and Yakir, B. (2000) 'Tail probabilities for the null distribution of scanning statistics', *Bernoulli*, 6(2): 191–213.

Snyder, J.P. (1987) *Map projections: A working manual*. Professional Paper 1395. Washington, DC: US Geological Survey. https://doi.org/10.3133/pp1395

Sokal, R.R. and Rohlf, F.J. (1981) *Biometry: The Principles and Practice of Statistics in Biological Research* (2nd edition). New York, NY: Freeman.

Stephens, M.A. (1974) 'EDF statistics for goodness of fit and some comparisons', *Journal of the American Statistical Association*, 69: 730–7.

Stewart, J.Q. and Warntz, W. (1958) 'Macrogeography and social science', *Geographical Review*, 48(2): 167–84.

Stone, R. (1988) 'Investigation of excess environmental risks around putative sources: Statistical problems and a proposed test', *Statistics in Medicine*, 7: 649–60.

Stouffer, S. (1940) 'Intervening opportunities: A theory relating mobility and distance', *American Sociological Review*, 5: 845–67.

Strahler, A.M. (1952) 'Hypsometric (area-altitude curve) analysis of erosional topography', *Geological Society of America Bulletin*, 63: 1117–42.

Tango, T. (1995) 'A class of tests for detecting "general" and "focused" clustering of rare diseases', *Statistics in Medicine*, 14: 2323–34.

Tobler, W. (1970) 'A computer movie simulating urban growth in the Detroit region', *Economic Geogra-phy*, 46 (Supplement): 234–40.

Turnbull, B.W., Iwano, E.J., Burnett, W.S., Howe, H.L., and Clark, L.C. (1990) 'Monitoring for clusters of disease: Application to leukemia incidence in upstate New York', *American Journal of Epidemiology*, 132: S136–43.

Waller, L.A. and Gotway, C.A. (2004) *Applied Spatial Statistics for Public Health Data*. New York, NY: Wiley.

Weissman, I. (2017) 'Sums of squares of uniform random variables', *Statistics and Probability Letters*, 129C: 147–54.

Weiszfeld, E. (1937) 'Sur le point pour lequel la somme des distances de n points donnes est minimum' (On the point for which the sum of the distances from given points is minimized), *Tohuku Mathe-matical Journal, First Series*, 43: 355–86.

Wilson, A.G. and Kirkby, M.J. (1975) *Mathematics for Geographers and Planners*. Oxford: Oxford University Press.

Worsley, K. (1996) 'The geometry of random images', *Chance*, 9(1): 27–40.

Yao, Y.-C. and Davis, R.A. (1986) 'The asymptotic behavior of the likelihood ratio statistic for testing a shift in mean in a sequence of independent normal variates', *Sankhya A*, 48: 339–53.

Yuill, R.S. (1971) 'The standard deviational ellipse: An updated tool for spatial description', *Geografiska Annaler: Series B, Human Geography*, 53(1): 28–39.

Index

www.ingramcontent.com/pod-product-compliance
Lightning Source LLC
Jackson TN
JSHW062325310126
97517JS00012B/119